Preface

From a small start in remote areas of the country, cable tv has grown to a major industry with thousands of systems in operation and hundreds of manufacturers supplying equipment. As the radio spectrum becomes more and more crowded, cable tv appears to be a practical way to make more spectrum space available. Considering that 600 separate 10-kHz two-way radio channels require about the same amount of spectrum space as a single tv signal, it seems that eventually all television will be carried on cables, and the completely "wired city" will become a reality.

Bringing good reception to areas that otherwise would have only marginal tv reception is still the major function of most cable tv systems. However, other applications are spreading rapidly. As the number of subscribers increases, the now secondary uses may well surpass the original applications.

In spite of the growth of cable tv in recent years, comparatively few books have been written on cable tv technology. This book is intended for the electronics engineer or technician who wishes to apply his knowledge of electronics to cable tv. In the first few chapters, the basic principles of television and transmission lines are reviewed, with particular emphasis on applications to cable transmission of broad-band signals. Later chapters cover the details of systems.

The author wishes to acknowledge the help and encouragement of all those who made the book possible. Manufacturers have been most generous in furnishing technical information and artwork; fellow members of the Cable Television Advisory Committee helped with technical details; and members of the Society of Cable Television Engineers gave freely of their knowledge. Particular mention must be made of the technical information supplied by Gary Arlen of the National Cable Television Association. Last, but certainly not least, I want to thank Grace, who has always been an inspiration to me.

JOHN E. CUNNINGHAM

Contents

CHAPTER 11

CHAPTER 12

CHAPTER 13

CHAPTER 14

CHAPTER 15

CHAPTER 16

CHAPTER 17

CHAPTER 18

CHAPTER 19

APPENDIX A

APPENDIX B

CHAPTER 1

Introduction to
Cable TV

Early cable tv systems were usually quite simple. Their only function was to provide television signals in areas where off-the-air reception was either unavailable or unsatisfactory. Signals from the cable were of poor quality by today's standards, but subscribers were not very discriminating. Almost any television picture was considered better than no picture at all.

In recent years, tv receivers have been improved considerably; subscribers have become more critical, and the Federal Communications Commission (FCC) has tightened its rules governing cable tv systems. As a result, new systems were designed and old systems updated to provide performance that was far superior to what was considered adequate only a few years ago.

Cable tv is no longer considered a substitute for off-the-air reception. Cable systems serve all types of communities, even those where several good-quality signals can be received on a modest home antenna system. It is being widely recognized that cable tv can not only provide more and better pictures in almost any location, but also offer much more than a few entertainment-type programs. Many cable systems originate their own programming, and an increasing number of systems have a two-way capability that opens the way to a wide variety of services that were not even dreamed of a few years ago.

Most of the signals that are carried by a cable tv system are picked up by high-gain antennas supported by tall towers, as shown in Fig. 1-1. In addition to the obvious function of providing the best possible reception of desired signals, the antenna has the equally important function of discriminating against undesired signals on the same or adjacent channels. This latter function is often the more critical design factor.

Television signals are then distributed to subscribers' homes through a coaxial-cable network. All of the signals will be attenuated in passing through the cable, with the highest frequencies suffering the most attenuation. To compensate for this loss of signal strength, amplifiers such as the one shown in Fig. 1-2 are inserted periodically along the cable. Unfortunately, in addition to amplifying the signals, each amplifier also introduces some noise and distortion. Inasmuch as amplifiers are cascaded one after another in the system, the noise and distortion are cumulative. As a result, there is a practical limit to the number of amplifiers that can be cascaded, and hence to the length of cable than can be used and still provide satis-

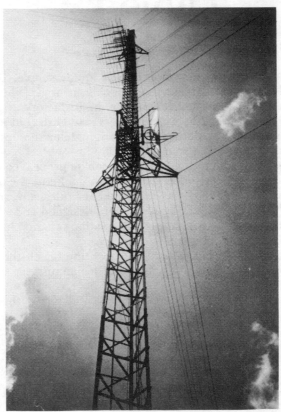

Courtesy National Cable Television Association

Fig. 1-1. A cable tv antenna system.

factory picture quality. The early systems were only a few miles long and used only a few amplifiers. In modern amplifiers, noise and distortion are minimized, so systems over 25 miles long with over 50 cascaded amplifiers are practical.

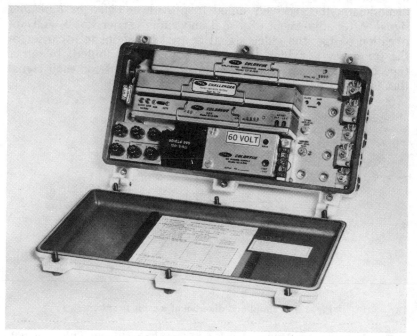

Fig. 1-2. A cable tv amplifier.

Fig. 1-3 shows a block diagram of a somewhat simplified cable tv system. The graphic symbols used in this diagram and throughout the book are those recommended by the National Cable Television Association (NCTA). A complete list of these symbols is given in Appendix A. The start of a cable tv system is called the *headend*. There, signals from television broadcast stations are picked up by the antennas, and in some systems distant stations are received via microwave links. At the headend, the signals are also processed for transmission by the cable network. This processing includes amplification, the setting of aural and visual carrier levels, the generation of pilot signals for automatic gain control (agc), and the combining of all the signals so they can be fed into a single cable. Almost invariably, ultrahigh-frequency (uhf) signals are converted to very high frequency (vhf) signals, and usually the vhf signals must be converted to a different channel to avoid interference.

There are two different types of signal-processing systems in common use. The simplest is the *heterodyne processor,* in which signals are first converted to a lower intermediate frequency and then converted back to the higher frequency at which they are to be distributed in the cable. Another, more complex, processor demodulates the received signal to provide a *video* or *baseband* signal. After adjustment of visual and audio carrier levels, and application of agc, the video signal is used to modulate a carrier of the frequency at which the signal is to be distributed. In either type of processor, a separate channel is provided for each signal that is to be carried.

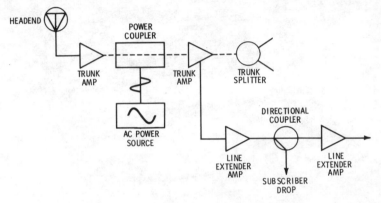

Fig. 1-3. A simplified diagram of a cable tv system.

The last item at the headend is a signal combiner. This is a network that allows all of the signals to be fed to the same cable with a minimum amount of interaction between channels.

The first section of cable leaving the headend is called the *trunk* cable. One or more trunk cables carry the signals along the main route or routes of the system. At intervals along the cable, *trunk amplifiers* are installed to compensate for the attenuation of the cable. The attenuation of the coaxial cable varies with the temperature, so it is necessary for some of the amplifiers to have temperature-compensation circuits that will keep the signal level nearly constant.

Connections to the trunk cable are made by means of *bridging* amplifiers. These amplifiers have a high input impedance and therefore will not disturb or load the trunk cable. The outputs of the bridging amplifiers are then fed to distribution lines that bring the signals to the subscriber's street. The distribution lines are also made of coaxial cable, but the cable is usually smaller than trunk cable. When the distribution lines are long, amplifiers called *line extenders* are installed to bring the signal to a satisfactory level.

Near the subscriber's home, a drop cable is connected to the distribution cable. Because of its short length and the fact that a flexible cable is easier to handle, a drop cable is usually a lower-grade, flexible cable.

Inside the subscriber's home, the drop cable usually terminates in a *balun transformer* that converts the unbalanced, 75-ohm impedance of the cable system to the balanced, 300-ohm input impedance of the television receiver. In many systems, the home equipment also includes a *set-top converter* that is used to permit reception of more than the usual 12 vhf channels. Fig. 1-4 shows a typical balun transformer of the type used at a subscriber's television set, and Fig. 1-5 shows a set-top converter.

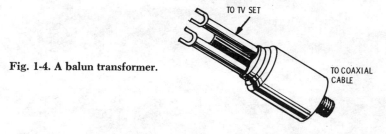

TO TV SET

TO COAXIAL CABLE

Fig. 1-4. A balun transformer.

FCC RULES

Cable tv began with practically no regulation. Some municipalities attempted to regulate their cable systems in a limited way, but no agency regulated cable tv as a whole. In 1962, the FCC asserted limited jurisdiction over cable tv, and in 1965, rules were established for systems that received signals by microwave. It was not until 1968 that the FCC established rules for all cable tv systems. Even then, the regulation was limited, since the authority of the FCC to regulate cable systems was being challenged in the courts. In June 1968, the Supreme Court affirmed the Commission's position. Since then, the FCC has established more-detailed rules intended to ensure the orderly development of cable tv and to anticipate public demands. Rules governing systems operating in the top 100 markets are more detailed than those governing systems in other areas. The FCC Rules and Regulations are subject to frequent changes; only a brief outline is given here.

The FCC distinguishes four different classes of cable tv channels:

Class I cable channels carry regular television broadcasts that have been picked up off the air at the headend, or delivered to the cable system by microwave or by direct connection to a television broadcast station.

Class II cable channels carry programs originated by the cable system itself—cablecasting.

Class III cable channels carry signals other than those that can be picked up on a standard television receiver. These could be television programs that are coded so that they can only be received with a special decoder. Such programs might be "pay tv" programs or perhaps programs intended for a select group of viewers such as medical doctors. These channels also carry nontelevision programs

Fig. 1-5. A set-top converter.

such as data, stock-market quotations, etc. Class III channels are used for "downstream" transmission, that is, transmission from the cable system to the subscriber.

Class IV cable channels are for "upstream" transmission, that is, transmission from the subscriber's terminal to some other part of the system. This is the so-called two-way capability, and it is covered in detail in a later chapter.

Certain other FCC rules will seriously influence the direction in which future cable systems will evolve. Chief among these are the following:

1. Systems operating in the major television markets must have an available bandwidth of at least 120 MHz throughout the system. This bandwidth is equivalent to 20 regular 6-MHz television broadcast channels. For each class-I channel that is utilized, the system must provide another 6-MHz channel for class II or class III use.
2. All systems in the major market must at least have a capability for handling nonvoice signals from the subscriber's terminal to other parts of the system.
3. All systems having over 3500 subscribers must be equipped to originate programming in addition to automated programming of time and weather.

CABLE CHANNEL ALLOCATION

All of the earlier, and many of the present-day, cable systems can carry a maximum of only 12 tv channels, operating in that portion of the spectrum normally occupied by vhf channels 2 through 13. In order to comply with the FCC reguations, these systems must expand their channel capacity.

One way of providing more than 12 channels is simply to run two coaxial cables throughout the system. Each cable would operate in the channel 2–13 portion of the system. The two cables would be completely independent, each having its own amplifiers. At the subscriber's terminal, a switch would allow him to select either cable, as shown in Fig. 1-6. Each channel would be identified by a channel number that would identify its frequency and by a letter that would indicate which cable was carrying the signal. Thus, a signal on channel 2, carried by cable A, would be identified as

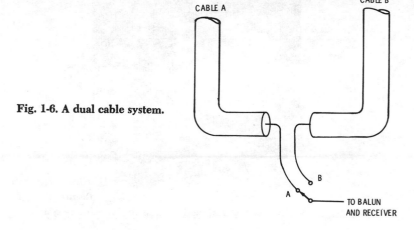

Fig. 1-6. A dual cable system.

channel 2-A, whereas a signal on channel 2 carried by cable B would be identified as channel 2-B.

The two-cable system is used extensively and in most areas will provide a capacity of about 20 usable channels. The reason that the full 24-channel capability cannot be used in many areas is that strong signals from local stations cause interference.

Another way to provide more than 12 channels, which is being used in many new systems, is to use a bandwidth ranging from a few megahertz to as high as 300 megahertz. When this is done, channel designations are assigned to the frequencies that are not normally designated as television frequencies. Table 1-1 shows the most commonly used channel designations. The use of this wide bandwidth is often called the *super-band* concept.

When the super-band concept is used, it is necessary to use some form of converter to convert the mid-band and super-band

Fig. 1-7. A color tv receiver designed specifically for cable tv systems.

Table 1-1. Cable TV Channel Designations

Channel Designation	Frequency (MHz)
Low Band	
2	54-60
3	60-66
4	66-72
5	76-82
6	82-88
Mid Band	
A	120-126
B	126-132
C	132-138
D	138-144
E	144-150
F	150-156
G	156-162
H	162-168
I	168-174
High Band	
7	174-180
8	180-186
9	186-192
10	192-198
11	198-204
12	204-210
13	210-216
Super Band	
J	216-222
K	222-228
L	228-234
M	234-240
N	240-246
O	246-252
P	252-258
Q	258-264
R	264-270
S	270-276
T	276-282
U	282-288
V	288-294

channels to channels that can be received on the subscriber's regular television receiver. This could be a converter located somewhere in the system, but it is usually a set-top converter at the subscriber's terminal.

Many of the problems encountered in modern cable tv systems carrying many channels can be traced directly to the receiver, rather than to the cable system. Most of the receivers in use were designed for operation in the typical home environment where the only signals received were those picked up by the home antenna. Such receivers were never meant to be able to cope with interfering signals on the same or adjacent channels. The FCC allocation rules separate cochannel stations by many miles.

At least one manufacturer has introduced a color receiver designed specifically for use on cable tv systems. Such a receiver is shown in Fig. 1-7. The small solid-state converter that the man is holding in his right hand is built into the tv set and replaces the set-top converter that he is holding in his left hand.

CHAPTER 2

Decibels, Signal Level, and Noise

Throughout a cable tv system the objective is to keep the signal level sufficiently higher than the noise level to provide good-quality pictures. In a cascaded transmission system, such as a cable system, it is customary to express gains, losses, and levels in decibels. There are several good reasons for this. First, the loss in a transmission line is exponential in nature and so is easily handled in terms of the decibel, which is a logarithmic unit. Second, when decibels are used, gains and losses along a cascaded system can be handled with simple addition and subtraction, without the need for complicated formulas involving multiplication and division. For these reasons, the technician should be familiar with decibel notation.

DECIBELS

As it was originally used, the decibel (dB) was a measure of a change in power level. Expressed in decibels, the change between two levels of power in a circuit is given by

$$N_{dB} = 10 \log \frac{P_1}{P_2} \qquad \text{(Eq 1)}$$

where,
N_{dB} is the number of decibels,
P_1 is one of the power values,
P_2 is the other power value.

For example, the power being dissipated in a resistor is 5 watts. If this power is increased to 10 watts, the change, in decibels, is given by

$$N_{dB} = 10 \log \frac{10}{5} = 10 \log 2 \qquad (Eq\ 2)$$

The logarithm of 2 is 0.301, so the change, in decibels, is

$$N_{dB} = 10 \log 2 = 10 \times 0.301 = 3\ dB \qquad (Eq\ 3)$$

It is worth remembering that

$$\log \frac{1}{x} = -\log x$$

This means that in Equation 1, we can always divide the larger power by the smaller. If the power is increasing, the sign of the change in decibels will be positive, and if it is decreasing, the sign will be negative. Thus in our example above, if the power were initially 10 watts and then decreased to 5 watts, the change would be *minus* 3 dB.

Although the decibel is a measure of change of power level, it can also be determined from voltage measurements, providing that the measurements are made across the same two points in a circuit, or across the same value of resistance. Our equation for the number of decibels change can be modified for using voltage instead of power by taking advantage of the fact that the power in a circuit is given by

$$P = \frac{E^2}{R}$$

where,
  E is the voltage,
  R is the resistance between the points where the voltage is measured.

Substituting E^2/R for P in Equation 1 gives

$$N_{dB} = 10 \log \frac{E_1^2/R_1}{E_2^2/R_2} = 20 \log \frac{E_1}{E_2} + 10 \log \frac{R_2}{R_1}$$

If the two values of R are the same, the second term on the right-hand side of this equation becomes zero, and the change in decibels is simply

$$N_{dB} = 20 \log \frac{E_1}{E_2}$$

Note that doubling the voltage causes twice the decibel change that doubling the power produces. This is because power is proportional to the square of the voltage. Doubling the voltage across a resistor will cause the power to increase four times. This point should be kept in mind when you are making decibel calculations.

It should be clear that the decibel, as we have explained it so far, is a measure of *change*, not of absolute level. In the above example, it was explained that saying a signal increased by 3 dB was equivalent to saying that the signal power doubled. Nothing is said or even implied about what the absolute value of power is. The power could change from 5 watts to 10 watts as in our example, or it could change from 500 to 1000 watts—in either case, the change in level is 3 dB.

dBmV

It is rather obvious that it would be desirable to have a unit like the decibel that could be used to measure signal level as well as changes in level. If we wish to use the decibel as a measure of signal level, we must select some arbitrary value of signal level which we will call zero dB. Since the impedance of most of the circuits used in cable tv is 75 ohms, the zero reference level has been selected as 1 millivolt measured across 75 ohms. When this reference is used, signal levels may be expressed in decibels above 1 millivolt. The symbol dBmV is used to represent this. A simple calculation shows that the reference power level of zero dBmV is 1/75 (0.0133) microwatt.

The equation for the number of decibels above one millivolt, or dBmV, in terms of power is

$$N_{dBmV} = 10 \log \frac{P}{1.3 \times 10^{-8}} \qquad (Eq\ 4)$$

and in terms of voltage is

$$N_{dBmV} = 20 \log \frac{E}{0.001} \qquad (Eq\ 5)$$

Equation 5 is based on the assumption that the voltage measurements are made across a resistance of 75 ohms. Thus, if the signal voltage across a 75-ohm cable were 2 millivolts, the signal level in dBmV would be

$$N_{dmBV} = 20 \log \frac{0.002}{0.001} = 20 \log 2 = 6 \text{ dBmV}$$

COMBINING DECIBELS

One of the advantages of using decibel notation is that calculations of gain and loss in a system involve only addition and subtraction, rather than multiplication and division. This is illustrated in Fig. 2-1. In Fig. 2-1A, the signal level is expressed in volts, and gains and losses are expressed as numeric ratios. In order to find the signal level at the output of the system, we must divide the voltage by the losses and multiply it by the gains as shown in the figure. By contrast, in Fig. 2-1B, the signal level is given in dBmV, and the gains and losses are given in dB. To find the output signal level in dBmV, we merely have to add the gains and subtract the losses. In a long cable tv system, this saves a considerable amount of labor.

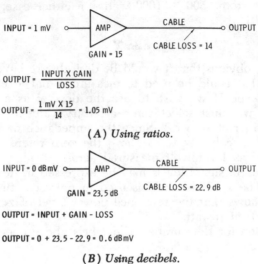

(A) Using ratios.

(B) Using decibels.

Fig. 2-1. Gain and loss calculations.

NOISE

There are several different types of noise that must be considered in a cable tv system. There is noise introduced by amplifiers, noise picked up by the antennas along with the desired signal, and interfering signals which must be considered as noise.

Perhaps the most basic, unavoidable source of noise is *thermal* noise, which is inherently present in all conductors. The noise power that could be delivered by any conductor to a matched load is given by the equation:

$$P = kTB$$

where,

k is 1.38×10^{-23} joules per kelvin. This is a constant from physics known as *Boltzmann's constant*.

T is the absolute temperature of the source expressed in kelvins, B is the bandwidth of the system in hertz.

The bandwidth of an ordinary television channel is about 4 MHz, and ordinary room temperature of 68°F is 293.2 kelvins. By substituting these values into our equation, we find that the thermal noise power in a television channel has a level of about 1.62×10^{-14} watt.

Substituting this value into Equation 4 gives the thermal noise level in dBmV as about −59.1 dBmV. This is the theoretical limit, or the lowest possible noise level that will be encountered in any physical system.

Fig. 2-2 shows a plot of thermal noise as a function of temperature. From this curve, it can be seen that the noise level does not change drastically with temperature, and for all practical purposes, we can consider the thermal noise level as −59.1 dBmV at all temperatures that are likely to be encountered in a cable television system.

The National Cable Television Association has defined the *system noise power* of a cable tv system as the noise power in watts that will be delivered by the system to a 75-ohm load over a bandwidth of 4 MHz. The *system noise level* is this power expressed in decibels, using a zero reference level of 1/75 microwatt (0 dBmV). In practice it is usually the system noise level, rather than the noise power, that is measured. The two quantities are related by Equation 4 on page 21.

Of course, the thermal noise described above is an ideal case that can only be attained in theory. All the amplifiers—from the preamplifiers in the headend to the amplifiers distributed throughout the system—introduce noise.

NOISE FACTOR AND NOISE FIGURE

The usual measure of the amount of noise that a stage contributes to a signal is called the *noise factor* or *noise figure*. The noise factor, F, is defined as the ratio of the signal-to-noise ratio at the input of the stage to that at the output of the stage; that is

$$F = \frac{S_i/N_i}{S_o/N_o}$$

where,
    N_i is the noise power at the input,
    N_o is the noise power at the output,
    S_i is the signal power at the input,
    S_o is the signal power at the output.

All the quantities are expressed in the same units of power, such as watts. This is an easy mathematical expression to understand, but it is difficult to work with. Calculations are made much easier by using the noise figure (F_N), which is the noise factor expressed in decibels:

$$F_N = 10 \log\frac{S_i}{N_o} - 10 \log \frac{S_o}{N_o}$$

where the signal and noise power are expressed in the same units. The noise figure is thus the amount in decibels that the signal-to-noise ratio is degraded in passing through the stage. For example,

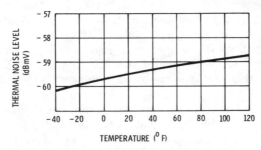

Fig. 2-2. Thermal noise level versus temperature.

if the signal-to-noise ratio at the input of a cable amplifier were 40 dB, and the amplifier had a noise figure of 6 dB, the signal-to-noise ratio at the output would be 34 dB.

GRAPHICAL ANALYSIS

The complete engineering design of a cable tv system requires many rather involved mathematical calculations. These calculations are simplified somewhat by the use of decibel notation, but the operations are nevertheless very tedious. Often, digital computers are used for this purpose. Although mathematical completeness is necessary in design, for purposes of understanding a system well enough to operate and maintain it, mathematical complexity is neither necessary nor desirable. Fortunately, graphical methods can be used to simplify the calculations considerably. A graphical presentation will not only adequately explain system operation, but also clearly display concepts that might seem obscure in mathematical expressions.

The graphical method can be simplified considerably by using graphs with coordinates that are well suited to the particular relationship being investigated. In many cases, selection of the proper type of graph paper will make what would otherwise be curves turn out to be simply straight lines. The various types of graph coordinates used in this book are explained in the following paragraphs.

Linear Graph Paper

Fig. 2-3 shows an ordinary linear graph in which the divisions of both coordinates are equally spaced. The line on the graph is a plot of the simple equation:

$$y = 2x$$

The y values are plotted on the vertical axis, and the x values on the horizontal axis. Because the graph is a straight line, the equation is called a *linear* equation by mathematicians. Linear graph paper is ideal for plotting such equations.

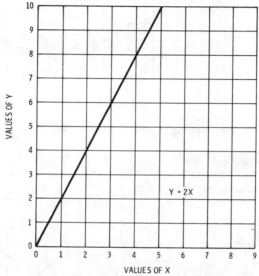

Fig. 2-3. A graph of a linear equation.

Now consider an equation such as the one given earlier for the number of decibels as a function of the ratio of two powers. Fig. 2-4 shows the graph of such an equation plotted on linear graph paper. It is obvious that the resulting graph is a curve that cannot be read with the same accuracy at all points on the graph. For example, it is easy to see that a power ratio of 2 to 1 represents a change of 3 dB, but the graph runs off the scale at ratios slightly greater than 10 to 1.

Semilogarithmic Graph Paper

Just as using a logarithmic quantity like the decibel simplifies mathematical calculations, using graph paper that has a logarithmic scale often simplifies plotting a graph. Fig. 2-5 is a plot of the same equation that was plotted in Fig. 2-4, but in this figure, the divisions

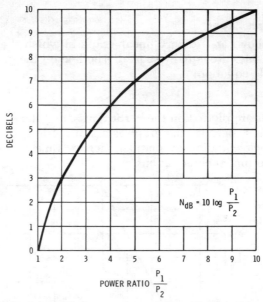

Fig. 2-4. Decibels vs power ratio on a linear graph.

along the horizontal axis of the graph are spaced logarithmically, like the divisions on a slide rule. The divisions along the vertical axis are spaced linearly, just as in Fig. 2-4. This type of graph paper is called *semilogarithmic,* or *semilog,* paper because one of the scales is logarithmic.

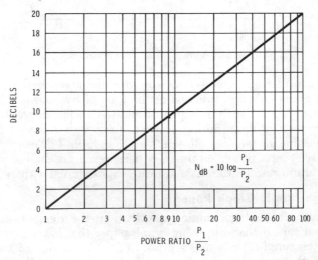

Fig. 2-5. Decibels vs power ratio on a semilogarithmic graph.

The result of using this type of graph is that the line that was a curve in Fig. 2-4 becomes a straight line in Fig. 2-5. This particular graph is called a *two-cycle* graph because two decades are encompassed along the horizontal axis. Another feature of the logarithmic scale is that it cannot go to zero. The lowest number must be some power of ten, such as 0.1, 1, or 10.

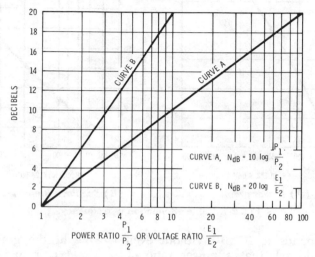

Fig. 2-6. Decibels vs power or voltage ratio.

Two things are obvious from an inspection of Fig. 2-5:

1. By using a horizontal scale with logarithmic divisions, a much larger range of power ratios can be covered than could be covered on a linear graph of the same size. There is no reason why we had to stop with two decades. We could have expanded the graph to three or more decades.
2. The same equation that produced a curved graph on linear graph paper produced a straight line on the semilogarithmic paper.

A more careful examination of Fig. 2-5 will disclose two other features that are not immediately obvious:

1. The graph can be read with the same *percentage* accuracy at all points. This was not possible in Fig. 2-4.
2. An equation of the form graphed varies at a constant number of decibels for a given percentage change in power ratio. Note that the level will change 3 dB every time that the power ratio is doubled or halved. For example, a power ratio of 2 corresponds to a change of 3 dB, and a power ratio of 4 cor-

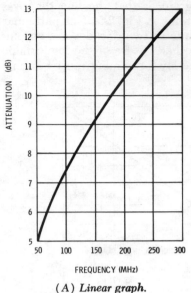

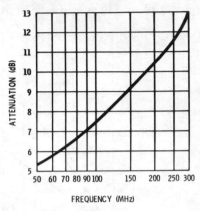

FREQUENCY (MHz)

(A) Linear graph.

FREQUENCY (MHz)

(B) Semilogarithmic graph.

Fig. 2-7. Cable attenuation vs frequency.

responds to 6 dB. Without even looking at the graph, we could tell that a power ratio of 8 would correspond to a change of 9 dB. Since the number of decibels changes by ten when the power ratio changes by a factor of ten, this curve is said to have a slope of 10 dB per decade.

Fig. 2-6 shows the same graph as in Fig. 2-5, together with another curve marked *B*, which is the number of decibels as a function of the voltage ratio. Here, the number of decibels changes by six when the voltage ratio is doubled. This can also be stated by saying that the graph has a slope of 20 dB per decade.

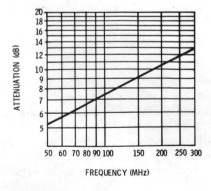

FREQUENCY (MHz)

Fig. 2-8. Cable attenuation vs frequency on a full logarithmic graph.

Logarithmic Graph Paper

There are situations in a cable tv system for which a graph will be curved even when it is plotted on semilogarithmic graph paper. One frequently encountered case involves the attenuation of coaxial cable as a function of frequency. For reasons that will be explained later, the attenuation of a given length of coaxial cable is not the same at all frequencies. In fact, the attenuation of a given length of cable is very nearly proportional to the square root of the frequency at which the attenuation is measured. A simplified version of the equation for cable attenuation has the form

$$L = K \sqrt{f}$$

where,

L is the attenuation in dB,
f is the frequency,
K is a constant of proportionality.

Fig. 2-7A shows a graph of attenuation plotted on linear graph paper, and Fig. 2-7B shows the same graph on semilogarithmic paper. In both cases the line is curved. In Fig. 2-8, we have the same graph, but this time it is plotted on graph paper on which the divisions on both axes are spaced logarithmically. Now the graph is a straight line. This type of graph paper is called *full logarithmic, logarithmic,* or *log-log* graph paper. There may be any convenient number of decades or cycles on either axis.

The examples show the principles of selecting a graph paper with coordinates that will simplify the graph itself. In many cases, the resulting graph becomes simply a straight line.

CHAPTER 3

The Television Signal

The objective of a cable tv system is to deliver high-grade television signals to subscriber's homes. The television signal itself is a complex waveform that contains information about the color and brightness of the scene, synchronizing information, and color-reference information. To reproduce the signal faithfully, all parts of it must be transmitted with a minimum amount of distortion. To understand the requirements of each element of a cable tv system, we must first understand the television signal itself and how each component of the signal affects the picture. Many faults in cable tv systems can be traced to an inadequate understanding of the signal and system requirements for faithful transmission.

There are many different ways in which a tv signal can be analyzed. Entire books have been devoted to the subject. The approach taken here is intended as a review of the principles of television, with particular emphasis on those aspects that are most significant in cable transmission.

VISUAL PERCEPTION

There are five aspects of the way in which the human eye perceives scenes that are of particular interest in television. These are *scene illumination, color, contrast, pattern recognition,* and *motion.* A discussion of color will be deferred until late in the chapter.

The *illumination,* or *brightness,* of a scene is simply the average brightness of the entire scene being viewed. It is the way in which a scene viewed in daylight differs from the same scene viewed under dim artificial light.

Contrast is the difference in light level from one part of a scene to another. The eye is capable of distinguishing many different levels of brightness in a scene. In order to reproduce a scene faithfully, a television system must be able to display several different light levels in a scene. In this way, light objects will appear to be light in the scene viewed on the television screen and dark objects will appear to be dark, just as they would be in the original scene. With insufficient contrast, all objects will be nearly the same—sort of gray.

Pattern recognition is the ability of the eye to distinguish the shapes of objects. It is related, but not linearly, to illumination. If the illumination is too low, the eye will not be able to distinguish the shapes of objects, particularly small objects.

When viewing a scene, the human eye perceives *motion* as a continuous process; that is, objects move smoothly from one place to another. It is the need to create a satisfactory illusion of motion that has had the greatest influence on the television system and the resulting waveforms.

The Illusion of Motion

The factors involved in creating the illusion of motion by displaying a series of still pictures have been studied in detail by the motion picture industry. Much of what was learned in connection with motion pictures has been applied directly to television. A brief review of the principles of motion picture projection will give the reader a better understanding of the standards adopted for television.

A motion picture film consists of a series of still pictures that are taken in rapid succession. In each frame, a moving object is displaced somewhat from its position in the preceding frame. When the pictures are displayed in rapid succession on a screen, the illusion of smooth motion is created. If the pictures, or frames, are displayed below a certain rate, the illusion of motion is still preserved, but the picture will appear to flicker. This flicker is very objectionable to the viewer. The rate at which the sensation of flicker becomes noticeable depends on the illumination. In general, lower rates can be tolerated at lower levels of illumination, but the relationship is not linear.

In a motion picture, the individual frames are displayed sequentially at a rate of 24 frames per second. This rate would normally produce objectionable flicker, but the sensation of flicker is overcome by the simple expedient of flashing each frame on the screen twice before advancing to the next frame. In television, frames are displayed at a rate of 30 frames per second and flicker is eliminated by a process called *interlacing*.

Scanning

It would be completely impractical for a television system to transmit an entire frame of a picture at one time. The only practical approach is to scan the scene and transmit each element of the picture separately.

The operation of a television camera tube is similar to that of a cathode-ray tube. The scene is focused on a photosensitive plate which is scanned by an electron beam. As shown in Fig. 3-1, the beam starts to scan the scene from the upper left-hand corner. The scanning motion is from left to right, and from top to bottom of the scene. (Because of the optical system, the actual scene may be upside down inside the camera, but this need not be considered here.)

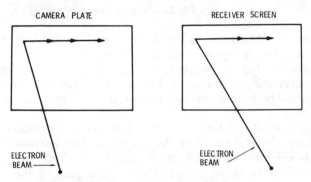

Fig. 3-1. Synchronization of camera and receiver.

As the electron beam sweeps across the plate, an electrical signal is developed that is proportional to the brightness of the scene at each particular spot. In order to reproduce the scene at the receiver, the electron beam in the picture tube must be synchronized exactly with the beam in the camera at all times. Every component in the entire television system, including all of the cable tv system, must maintain this synchronization.

There are actually two scanning systems in a television camera or receiver. The horizontal system sweeps the electron beam smoothly from left to right at a constant speed, then rapidly back to the left side. At the same time, the vertical scanning system is moving the beam from the top of the frame to the bottom at a much slower rate. This is illustrated in Fig. 3-2, which shows a few scanning lines. Since the beam is being deflected to the right rapidly, and toward the bottom at a lower rate, each line slopes slightly downward as shown. At the end of each scanning line, the beam is swept rapidly to the left as shown by the dashed lines

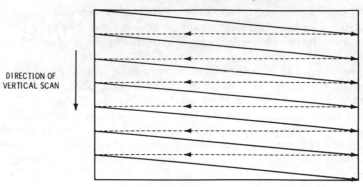

DIRECTION OF
VERTICAL SCAN

Fig. 3-2. Tv scanning lines.

in the figure. When the beam finally reaches the bottom of the picture, it is swept rapidly back to the top.

It is obvious that the beams in the camera and in the picture tube of the receiver must start each line at the same time. Signals that synchronize the two beams are transmitted along with the picture information.

During the *retrace intervals*—the times when the beam is being swept from the right side of the screen to the left and from the bottom to the top—the picture tube must be blanked out. This is accomplished by blanking pulses which are a part of the television waveform.

The television picture is thus made up of a number of horizontal lines. The number of lines that make up one frame depends on the ratio of the horizontal scanning frequency to the vertical scanning frequency. This can be seen from the fact that if the two frequencies were the same there would be only one scanning line extending from the upper left-hand corner of the picture to the lower right-hand corner.

It is obvious that the resolution—the number of separate picture elements that can be resolved—will be greater if the number of horizontal scanning lines is increased. Unfortunately, the bandwidth required for transmission will also be increased. The standard adopted in North America calls for 525 lines per frame. The horizontal lines are 1/3 longer than the height of the picture, giving an aspect ratio of 4 to 3, which is also standard for motion pictures.

Interlacing

The number of lines per frame, the vertical and horizontal scanning frequencies, and the bandwidth required for transmission are all related. In order to conserve bandwidth, the frame rate is kept as low as practical. A rate of 30 frames per second is adequate to create the illusion of motion, but when viewed under the lighting conditions of television, it would result in an objectionable amount of flicker.

The North American standard actually uses a 30-frame-per-second rate, and the flicker problem is solved by an ingenious arrangement called interlacing.

Interlacing is accomplished by using horizontal and vertical scanning frequencies that are not integrally related. The horizontal scanning frequency is 15,750 lines per second, and the vertical frequency is 60 Hz. Using simple division we find that 15,750 divided by 60 equals 262½, which means that there will be 262½ horizontal lines displayed in the time required for the beam to travel from the top of the screen to the bottom. If the beam starts at the upper left-hand corner, there will be 262 full lines, and half of another line will be displayed when the beam reaches the bottom of the screen. That is, the beam will be at the middle of the screen when it reaches the bottom. This group of 262½ lines, which represents half of a frame, is called a *field*. At this point, the beam is swept to the top of the screen and another field of 262½ lines is transmitted starting at the middle of the screen. In this way, the lines of the second field are *interlaced* between the lines of the first field. This is shown in Fig. 3-3. The interlacing eliminates the flicker effect in much the same way that flashing the same picture on the screen twice eliminates flicker in motion pictures.

Resolution

For a television picture to appear realistic and to be acceptable to the viewer, there must be a lot of detail. A good television picture must show at least 100,000 separate picture elements. The amount

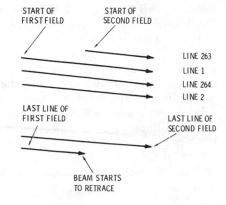

Fig. 3-3. Principle of interlacing.

of detail that is contained in a picture is measured in terms of its *resolution.*

Resolution is a term that has been borrowed from the motion picture industry where it is defined somewhat differently than it is in television. To avoid confusion, the meaning of the term resolution as it is used in television should be clearly understood.

Picture resolution is defined in television as the number of separate vertical or horizontal lines that can be discerned in the picture. Several factors influence the resolution of a television system, including the size and shape of the electron beams in the camera and picture tubes, and the overall bandwidth of the system. Because the television picture is made up of horizontal lines, horizontal and vertical resolution must be considered separately.

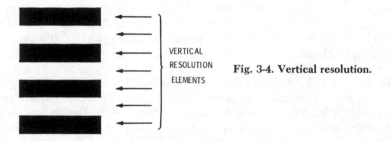

VERTICAL RESOLUTION ELEMENTS

Fig. 3-4. Vertical resolution.

Vertical resolution is the number of separate horizontal picture elements that can be discerned along a vertical line in the picture as shown in Fig. 3-4. It is limited by the number of horizontal lines used to make up the picture. Superficially, it might appear that in a 525-line system, 525 separate picture elements might be discerned along a vertical line. In actuality, this is not true. In the first place, as was pointed out above, it is necessary to transmit blanking and synchronizing information along with the picture information. In practice, this takes about 35 scanning lines, leaving a maximum of 490 lines for the picture itself. Actually, even a 490-line resolution is not attainable in a 525-line system.

Fig. 3-5 shows part of a pattern consisting of 490 lines which is to be televised. In Fig. 3-5A, the electron beam falls squarely in the middle of each of the lines, and the 490-line pattern is transmitted faithfully. In Fig. 3-5B, however, the electron beam straddles the edge of each line so that it sees both the line and the white space between lines. Since the electron beam can have only one value at each point, it will have a value corresponding to the average illumination of the line and the white space. The result is that the picture will be gray with no evidence whatsoever of the pattern that is being scanned.

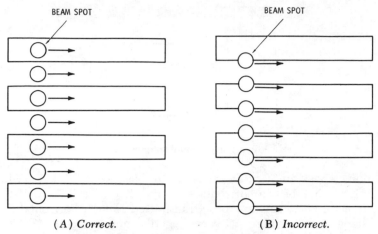

(A) Correct. (B) Incorrect.

Fig. 3-5. Effect of beam on resolution.

The example of Fig. 3-5 is extreme and would not be encountered in practice. Experiments have shown, however, that only about 70% of the scanning lines are actually effective, giving us $490 \times 0.70 = 350$ lines of vertical resolution. This is about the best value of vertical resolution that will be encountered in practice.

Horizontal resolution is defined as the number of vertical picture elements that can be discerned along a horizontal line in the picture as shown in Fig. 3-6. Since the horizontal lines are continuous, horizontal resolution is influenced by factors other than those which influence vertical resolution. In the interest of a balanced picture, the vertical and horizontal resolution should be about the same. Since the horizontal lines are 1/3 longer than the height of the picture, the horizontal resolution should be about $4/3 \times 350 = 466$ lines.

Horizontal resolution is limited primarily by the bandwidth of the system, that is, by the number of times that the intensity of the

ELEMENTS OF
HORIZONTAL
RESOLUTION

Fig. 3-6. Horizontal resolution.

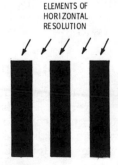

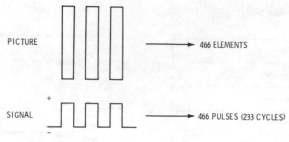

Fig. 3-7. Horizontal frequency.

scanning spot can vary from white to black and back to white during the time of one horizontal scanning line. Fig. 3-7 shows that if 466 picture elements are to be transmitted along a horizontal line, the signal must change between negative and positive 466 times during the time of one horizontal scanning line. Since the signal does this twice during each complete cycle, 233 cycles are required during each horizontal line.

Since the horizontal scanning frequency is 15,750 Hz, the time it takes for one horizontal line to go from the left side of the screen to the right side and back again is

$$\frac{1}{15,750} = 0.0000635 \text{ second}$$

or 63.5 microseconds. As with the vertical scanning signal, some time is required for transmitting synchronizing and blanking information. Actually, about 17% of the time is used for this purpose, leaving about 52.5 microseconds for actual picture information. Dividing this time into the number of cycles required gives

$$\frac{233}{0.0000525} = 4,438,000 \text{ Hz, or } 4.438 \text{ MHz}$$

which is the bandwidth required for 466 lines of horizontal resolution.

Usually, other factors also tend to limit horizontal resolution, and a figure of about 450 lines is common. Even at this resolution, each picture will consist of over 150,000 separate picture elements.

THE VIDEO WAVEFORM

The actual waveform required to transmit a scanned picture together with vertical and horizontal synchronizing and blanking information is quite complex. It is a combination of pulses and smoothly varying signals, and in general, all parts of the waveform must be transmitted faithfully for good picture quality. Fig. 3-8 shows a single line of a monochrome video signal. In this figure, a positive signal represents black and a zero signal level represents white. This is the polarity that is applied to the modulator of a television transmitter. However, at other places in a system, such as in the headend of a cable tv system, the signal may be inverted.

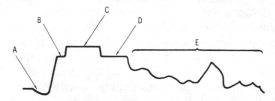

Fig. 3-8. Waveform of one horizontal line.

The first portion of the video signal, A (Fig. 3-8), is the video (picture) signal at the end of the previous line. The next portion, B, is called the *front porch*. It is 1.59 microseconds in duration, and its purpose is to blank the screen of the television receiver just before the beam starts to retrace from the right-hand side of the screen to the left. The front porch also tends to isolate picture information from synchronizing information. It is followed by the *horizontal sync pulse*, or *sync tip*, C, which is 4.76 microseconds in duration. The leading edge of the sync pulse is used to synchronize the sweep circuits in the receiver with those in the camera. The following portion, D, is called the *back porch*. It is 4.76 microseconds in duration, and its function is to keep the screen of the receiver blanked until the next line of picture information is transmitted. Finally comes the actual picture information, E, in which high levels correspond to black, and low levels correspond to white.

Fig. 3-9 shows the same waveform with relative levels indicated. It is common to express times in television signals as a function of H, which is the time elapsed from the start of one line to the start of the next line—63.5 microseconds.

A series of the waveforms—horizontal lines—shown in Fig. 3-9 will bring the beam in the receiver to the bottom of the screen. It is now necessary to blank the screen during the time required for the beam to return to the top of the screen, and to transmit a

Fig. 3-9. Scanning line.

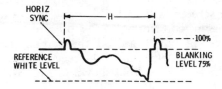

signal that will ensure vertical synchronization. During this period horizontal synchronization must be maintained. All this is accomplished by the family of field pulses shown in Fig. 3-10. The time scale of Fig. 3-10 is much longer than that of Fig. 3-9, which shows only one line.

The first part of the field pulses consists of six equalizing pulses, which will be discussed later in more detail. These are followed by a serrated vertical synchronizing pulse which, in turn, is followed by six more equalizing pulses. Throughout this entire interval, the pulses and the serrations of the vertical pulse maintain the spacing of the horizontal synchronizing pulses. Thus, horizontal synchronization is maintained throughout the interval when the beam is returning from the bottom of the screen to the top.

The synchronization process is more readily understood if we remember that horizontal synchronizing information is contained in the leading edge of all the pulses, whereas vertical synchronizing information is contained in the duration of the serrated vertical sync pulse.

The function of the equalizing pulses preceding and following the serrated vertical sync pulse is often not fully understood. They are used to preserve proper interlacing. This is important in cable tv systems that originate their own programs.

It was mentioned earlier that the vertical sync information was obtained from the duration of the vertical sync pulse. In a receiver, this is detected by an integrating circuit that initiates the vertical sweep when the charge on a capacitor reaches a certain value. The capacitor is charged by the horizontal pulses as well as by the vertical pulse, but the amount of charge that accumulates during a horizontal interval is not great enough to trigger the vertical sweep circuit. It leaks off during the time of one line. A little reflection will disclose that inasmuch as one field ends with a complete line and the following field ends with half a line, there is a different time interval in alternate fields between the last horizontal pulse and the beginning of the vertical pulse. If there were no equalizing pulses, one field would start at a slightly different time than the following field. This would result in the scanning lines in alternate fields being separated unequally. The scanning would tend to "pair," thus degrading the quality of the picture.

Fig. 3-11 shows an outline of a television picture including the blanking and synchronizing pulses. Since an increasing signal makes the screen darker, the blanking pulses are dark when the picture is transmitted. Normally the picture is made large enough that the blanking pulses are off the screen. However, if the vertical and horizontal deflection of the receiver were reduced, the picture would, in fact, look like that shown in Fig. 3-11.

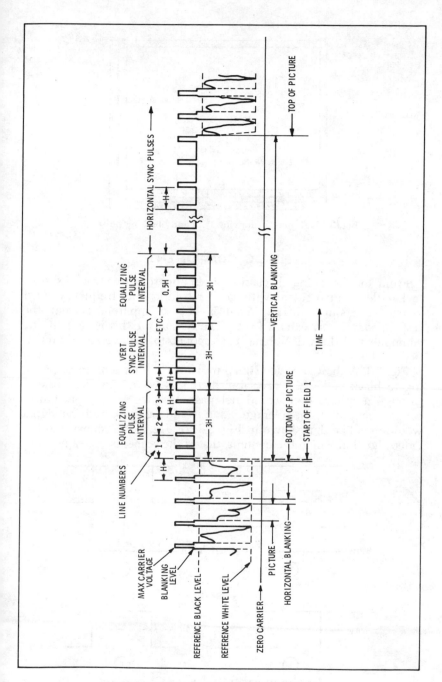

Fig. 3-10. Vertical synchronizing pulses.

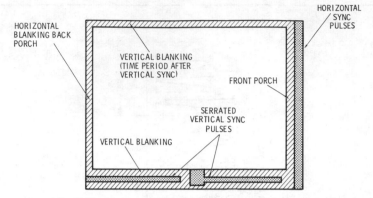

Fig. 3-11. A picture showing sync and blanking pulses.

THE DC COMPONENT

From the foregoing discussion, it is apparent that for a system to handle a video signal with good resolution, its frequency must extend to about 4 MHz. What is not so apparent is that the response must also extend to dc. The reason for this is that the dc component of the video signal corresponds to the average illumination of the scene.

Fig. 3-12A shows a video signal in which the portions corresponding to black and white are labeled. If this waveform were passed through a system that would not pass dc, such as a capacitance-coupled amplifier, the average value would be zero, and the signal would look like that shown in Fig. 3-12B. Fig. 3-12C shows another video signal in which the content, that is the scene, is the same as in

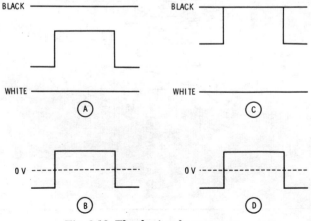

Fig. 3-12. The dc signal component.

Fig. 3-12A, but the average value of illumination is different. When this signal passes through a system that will not pass dc, the waveform is as shown in Fig. 3-12D. Note that parts B and D of Fig. 3-12 are identical and that all information about the average value of scene illumination or background illumination has been lost.

A circuit that will reconstruct the dc component of a video signal is called a *dc restorer*. Such circuits are used in the headends of some cable tv systems. At one time, all tv receivers had dc restorer circuits. In such a receiver the screen goes black between scenes. Unfortunately, when this happens, many viewers turn up the brightness control, upsetting the adjustment of the receiver. As a result, many manufacturers of monochrome receivers omit the dc restorer completely and let the viewer determine the average illumination with the brightness control on the receiver.

THE TRANSMISSION OF COLOR

A color television signal contains much more information than a monochrome signal, yet the bandwidth required for transmission is no greater. As a result, the system requirements for faithful color transmission are much more stringent than they are for monochrome signals. This is particularly apparent in cable tv systems that provide good monochrome picture quality, but provide color pictures that are barely acceptable.

The NTSC (National Television System Committee) system of color television used in North America was developed as a means of providing color television with a minimum amount of disturbance to the existing monochrome system. There have been many comments to the effect that since the NTSC system was a compromise, it is not capable of fine picture quality. This is not true. When all of the components of a television system, including the cable tv system, are properly adjusted, the system can provide excellent picture quality.

The principal consideration in the development of the NTSC color television standards was compatibility with the existing television system. Compatibility means that:

1. It must be possible to receive color broadcasts on monochrome receivers and to display them in black and white.
2. It must be possible to receive monochrome broadcasts on color receivers and display them in black and white.

Although there are many possible ways to transmit color television broadcasts, the requirements for compatibility rule out many of them. The system that is now used is completely compatible; furthermore, it permits transmitting color broadcasts with no increase in bandwidth. To facilitate understanding the color signal, we will briefly review some of the basic principles of color perception.

The Perception of Color

The sensation known as color is the result of the response of the eye to three different properties of light—*brightness, hue,* and *saturation.*

Brightness, or luminance, is related to the amount of light reaching the eye. If a large amount of reflected light from an object reaches the eye, the color is said to be bright; if only a small amount of light is involved, the color is said to be dim. Hue, often referred to as tint, is related to the wavelength of the light. It is the property of light that distinguishes one color from another. Red differs from blue in that it has a different hue.

Saturation determines how much a color differs from white. A pale color such as a pastel pink is low in saturation, whereas a deep red is highly saturated. There is more white in a lightly saturated color.

Practically any color, including white, can be produced by combining the proper amounts of three *primary* colors. This property of color makes it possible to re-create any desired color by transmitting only the proper amounts of the three primary colors. In the NTSC system, the three primary colors are red, green, and blue.

The human eye can detect small changes in brightness from one part of a scene to another, but it is not nearly as adept at distinguishing small changes in color. This is increasingly true as the colored area of the scene becomes smaller. This characteristic of human vision is of great importance in television. It permits the bandwidth of the color portion of the signal to be reduced considerably.

Another characteristic of vision that has a significant effect on the operation of a color television system is called *color adaptation.* This is the ability of the eye to adapt to the color of ambient light. When an observer comes out of bright sunshine into a room lighted by incandescent lamps, white objects will appear yellow, but after a very short time they will appear to be white again. Over a wide range, the eye accepts ambient light of almost any color as white, and it does a good job of perceiving what the actual colors of objects are.

Color adaptation affects the way that an observer will judge the quality of color of a television scene. Suppose, for example, that a scene is being televised outdoors under bright sunlight. If a viewer is watching the scene on a television set in a room lighted by incandescent light, the whites on the tv screen will appear decidedly blue.

The effect of color adaption on the design and operation of tv systems will be discussed later.

The Color System

From the above discussion it can be seen that one way of transmitting the chromatic content of a televised scene is to transmit signals corresponding to the brightness, hue, and saturation of each element of the picture. The NTSC system does exactly that, in a way that is compatible with the monochrome system described earlier and without increasing the required bandwidth.

Fig. 3-13 shows a sketch of a color tv camera. It actually consists of three monochrome cameras all looking at the scene through the same lens. The color of the scene is divided into red, green, and blue light by *dichroic* mirrors. A dichroic mirror is a plate of glass coated with a thin metallic layer. It has the property of reflecting one of the primary colors while allowing the others to pass through it. The camera has three outputs corresponding to the three primary colors.

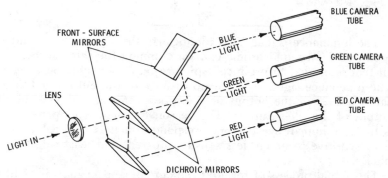

Fig. 3-13. Color separator system of a color tv camera.

A scene with the various colors marked is shown at A in Fig. 3-14. Parts B, C, and D of Fig. 3-14 show the outputs of the red, green, and blue cameras as the scene is scanned. The camera and the picture tube in the receiver are actually the only places in the entire system where signals corresponding to the primary colors occur. The picture is transmitted by converting these signals into other signals that correspond to the brightness, hue, and saturation of the scene. The brightness signal is usually called the *luminance* signal and is represented by the letter Y. The signals corresponding to hue and saturation are called *chrominance* signals. The process of combining the outputs of the three cameras to form chrominance signals is called *encoding* and is accomplished in a circuit called a *matrix*.

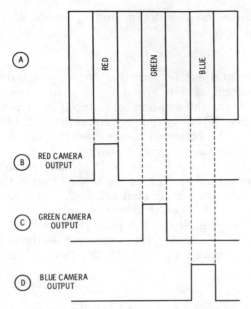

(A)

RED   GREEN   BLUE

(B) RED CAMERA
OUTPUT

(C) GREEN CAMERA
OUTPUT

(D) BLUE CAMERA
OUTPUT

Fig. 3-14. Color signals.

The luminance signal, Y, is identical to the signal that would be produced by a monochrome camera scanning the same scene. It is all that is needed to produce a picture on a monochrome receiver. The luminance signal is made up of 30% of the output of the red camera, 59% of the output of the green camera, and 11% of the output of the blue camera. The reason for these odd percentages is that the human eye does not respond equally to all colors. It is most sensitive to green, less sensitive to red, and least sensitive to blue.

The actual process of forming the chrominance signals is of little interest in cable tv. What is of interest is the nature of the resulting rf signal.

The luminance signal is used to amplitude-modulate the transmitter. Thus, it appears in the composite waveform just as in the monochrome waveform discussed earlier. The chrominance information must be transmitted in some other way. This is done by using two chrominance signals from the encoder to modulate a subcarrier in both amplitude and phase. The subcarrier itself is suppressed, only the sidebands being transmitted. The frequency of the subcarrier was selected so that it could be added to the signal without increasing the bandwidth required for transmission. The way this is done is described in the following paragraphs.

A spectrum analysis of a monochrome television signal would show that most of the picture information is not distributed evenly throughout the available bandwidth, but is clustered in bunches that

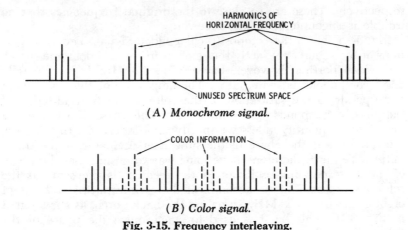

(A) *Monochrome signal.*

(B) *Color signal.*

Fig. 3-15. Frequency interleaving.

are separated by an amount equal to the horizontal line frequency, as shown in Fig. 3-15A. Because of this clustering, only about 54% of the available bandwidth is actually used. If the frequency of the color subcarrier is chosen to be an odd multiple of one half of the horizontal line frequency, the color modulation products will fall exactly between the clusters of the monochrome signal, as shown in Fig. 3-15B.

Several other factors must be considered in selecting a subcarrier frequency. For example, the subcarrier frequency must be high enough that the color variations will not show up as noise on the luminance signal. Another consideration is that the color subcarrier must not cause objectionable beats with the sound carrier.

As a result of all these considerations, a subcarrier frequency of 3.579545 MHz was finally chosen. This is not an exact multiple of one half the original monochrome horizontal line frequency, so the horizontal and vertical scanning frequencies were changed slightly for color transmission. The horizontal and vertical scanning frequencies used for color broadcasts are 15,734.26 Hz and 59.94 Hz,

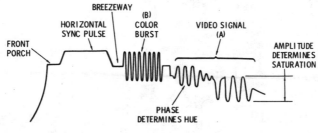

Fig. 3-16. Color tv waveform.

respectively. These are so close to the original frequencies that no trouble is encountered in the receiver.

Fig. 3-16 shows the composite color video signal. The sync portions of the signal are exactly the same as in the monochrome signal. The video portion, however, has a 3.579545-MHz (approximately 3.58) component as shown at A. The amplitude of this component corresponds to the saturation of the color being transmitted, and the phase corresponds to the hue or tint. The phase of a signal is a meaningless quantity unless we specify some reference. In this case, the reference is the phase of the subcarrier oscillator at the transmitter. We must, therefore, have some way to ensure that the phase of the corresponding oscillator in the receiver is the same as the reference phase. This is accomplished by transmitting a *burst* of eight cycles of a 3.58-MHz signal on the back porch, as shown at B in Fig. 3-16. This burst is used to synchronize the phase of the subcarrier oscillator in the receiver with the reference phase.

THE RF TELEVISION SIGNAL

In a cable tv system, we rarely encounter the video waveform itself, except in some headends and in the local studio. The signal that is normally encountered is the rf signal, which is modulated by the video signal. Fig. 3-17 shows the waveform of an rf television signal. The waveshape is just about what one would expect— a high-frequency carrier that is amplitude modulated by the video signal that we have been discussing. Fig. 3-18 shows the rf spectrum of this signal. This is not what we would expect from an amplitude-modulated signal since most of the information is transmitted on one side of the carrier.

The video signal in the studio has a bandwidth of about 4 MHz. If this signal were amplitude-modulated onto a carrier in the conventional way, the resulting rf signal would have a bandwidth of at

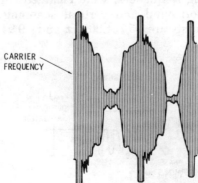

CARRIER
FREQUENCY

Fig. 3-17. Rf waveform.

least 8 MHz. This would severely limit the number of television stations that could be allocated in the available portion of the spectrum.

Fortunately, it is possible to recover all of the information from an a-m signal by using only the carrier and one sideband. It is possible, therefore, to cut the bandwidth of the signal in half. However, doing this would require very elaborate filtering arrangements, and in practice a compromise is made. Most, but not all, of the lower sideband is suppressed. This system, called *vestigial sideband* transmission, permits the bandwidth to be conserved without extremely elaborate filtering arrangements.

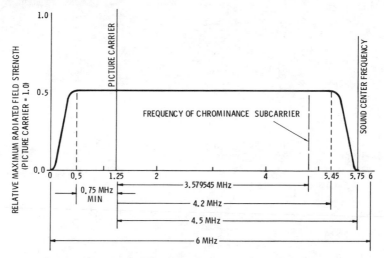

Fig. 3-18. Spectrum of a television signal.

As shown in Fig. 3-18, the television channel is 6 MHz wide. The video carrier is located 1.25 MHz above the lower edge of the channel. The spectrum space below the video carrier is occupied by the vestigial lower sideband. The upper sideband has a bandwidth of about 4 MHz. The color information is clustered near 3.58 MHz.

The sound portion of the broadcast is transmitted on a separate carrier that is 4.5 MHz above the video carrier. The sound carrier is frequency-modulated, with a deviation of ±25 kHz corresponding to 100% modulation. Federal Communications Commission regulations require that the video carrier frequency be maintained within 1000 Hz of its assigned value and that the difference between the sound and video carriers be held within 1000 Hz of 4.5 MHz.

CHAPTER 4

System Requirements

As described in the preceding chapter, the television signal is a rather complex waveform that contains picture, color, synchronizing, and blanking information. Ideally, any system that is intended to transmit television signals should faithfully reproduce at its output the complete television waveform with no distortion. No practical system can do this. Every element of the system will degrade the signal in some way, however slight. The object of proper system design, operation, and maintenance is to minimize the degradation of the signal. To accomplish this, we must be familiar with the ways that various system parameters will affect the television signal.

Of course, picture quality is really a subjective quantity. Measurements will determine how much a waveform has been distorted or degraded by noise, but whether or not a television picture is satisfactory depends on the response of the viewer. A picture is excellent, fair, or poor when the viewer judges it to be excellent, fair, or poor. Fortunately, a great deal of research has been done on viewer response in connection with television broadcasting. Most of this work can be applied directly to cable tv.

Fig. 4-1 shows a complete television system, including the cable link. The only place where the video waveform appears at video frequencies is in the studio and in the subscriber's receiver. At all other points, it is the modulated rf signal that is amplified and distributed. Nevertheless, it is useful to study the ways that the properties of a system can affect the video signal. In general, the same requirements will hold for the modulated rf signal, with the excep-

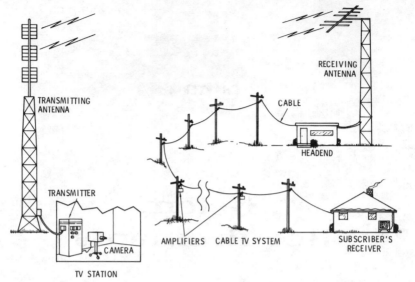

Fig. 4-1. Each component of entire system must faithfully reproduce the signal.

tion that the frequency range of interest centers around the rf carrier frequency rather than in the video range.

FREQUENCY RESPONSE

A television channel is 6 MHz wide. The picture, synchronizing, and color information occupy at least 4 MHz of this bandwidth. It is obvious, therefore, that any system that handles television signals must have a wide bandwidth. An exact analysis of how the frequency response of a system will affect a video signal is rather complex. Fortunately, it is not necessary for our purposes. We can study the effect of frequency response by using a simplified signal that represents the worst type of situation that we will encounter in practice.

Fig. 4-2A shows a simple picture that consists of only a single black bar against a white background. The corresponding waveform of one scanning line is shown in Fig. 4-2B. The signal is at zero level as the scanning proceeds from left to right until the bar is reached. The signal then rises rapidly to a positive level, which it holds until the edge of the bar is reached, when it drops rapidly to zero. Thus, each horizontal line produces a signal that is a single rectangular pulse. It might appear intuitively obvious that passing a sharply rising signal is equivalent to passing a high-frequency signal. This is true. In fact, it can be shown mathematically that any repeating wave-

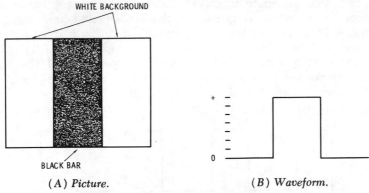

(A) Picture. (B) Waveform.

Fig. 4-2. Elementary picture and resulting waveform.

form, no matter what shape, can be represented by a sinusoidal signal together with its many harmonics. We will not present the mathematical treatment here, but will merely show the principle.

Fig. 4-3A shows a pulse train similar to the pulse of Fig. 4-2. Fig. 4-3B shows a sinusoidal signal having the same period. Admittedly, the signals do not look very much alike. In Fig. 4-3C we have the same sinusoidal signal, but this time it has small amounts of its third and fifth harmonics added to it. It is beginning to look more like the pulse signal. In Fig. 4-3D, we have the same sinusoid

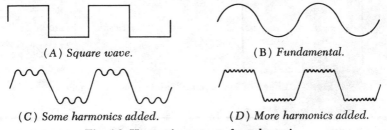

(A) Square wave. (B) Fundamental.

(C) Some harmonics added. (D) More harmonics added.

Fig. 4-3. Harmonic content of a pulse train.

with many harmonics. The resemblance to the pulse signal is even greater. From this figure we can see that if we added enough harmonics in the proper amount, we could in fact duplicate the pulse signal. Theoretically, to get the pulse signal with its straight sides and sharp corners would require an infinite number of harmonics. This tells us that to transmit our pulse signal with perfect fidelity would require a system having infinite bandwidth.

Of course, no system could actually have an infinite bandwidth, but neither could an actual television camera produce a perfectly rectangular pulse. In practice, a rectangular pulse signal can be handled with adequate fidelity by a system having a bandwidth of about ten times the frequency of the fundamental component.

Fig. 4-4A shows another pulse train. In order to transmit the rapidly varying parts of the pulse, we must transmit its high-frequency components. Thus, we might suspect that if the high-frequency response of a system is inadequate, the sides of the pulse will be more sloped and the corners rounded. This is true, and the effect is shown in Fig. 4-4B. The effect that this type of distortion

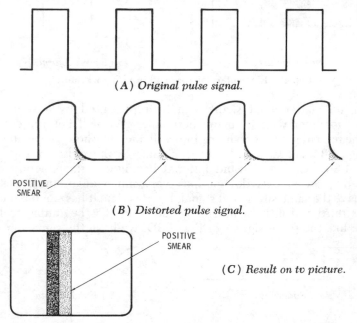

(A) Original pulse signal.

POSITIVE
SMEAR

(B) Distorted pulse signal.

POSITIVE
SMEAR

(C) Result on tv picture.

Fig. 4-4. Effects of inadequate high-frequency response.

has on the picture is shown in Fig. 4-4C. Note that there are no rapid transitions from white to black and vice versa. The transitions tend to be smeared. This type of smear is called *positive* smear because it is the same color as the picture element it follows—black smear follows a black picture element.

Inasmuch as the high-frequency response of the system affects primarily the rapidly varying parts of the signal, we might correctly suspect that the low-frequency response will affect the slowly varying parts of the signal. This is shown in Fig. 4-5. Fig. 4-5A shows the input waveform. In Fig. 4-5B we see the rapidly varying parts of the signal are all right because they are affected by the high-frequency response. The flat top of the wave, however, will tend to droop. This means that the right-hand side of the black bar will be lighter. Another effect of inadequate low-frequency response is that the signal will tend to "ring," or oscillate, at the end of a rapid transition. This again will cause smear in the picture, but in this case it will be negative smear. That is, the smear will be opposite from the picture element it follows. A white smear will follow a black picture element.

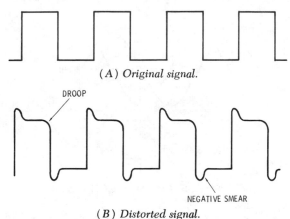

(A) Original signal.

(B) Distorted signal.

Fig. 4-5. Effect of inadequate low-frequency response.

Still another effect of inadequate frequency response involves color television signals. The luminance portion of the signal has a bandwidth of about 4 MHz. With a monochrome signal, if the high-frequency response is inadequate, the result will be some loss of picture detail or resolution. The amount of resolution that is lost will depend on just how bad the high-frequency response is. In many cases it might not be very objectionable. With a color signal, however, all of the color information is clustered about a frequency 3.58 MHz above the video carrier. If the system does not pass this frequency, the color in the picture will be "washed out" or, in an extreme case, completely absent. This effect can be simulated on a television receiver by adjusting the fine tuning until the color disappears and some of the picture detail is lost.

PHASE DISTORTION

The way in which the amplitude response of a system affects the signal and the resulting picture is straightforward. The effect of the phase response of a system is more subtle. Fig. 4-6 shows two sinusoidal signals having a frequency of 250 Hz and a phase difference of 90 degrees. One cycle of either of these signals has a time duration of $t = 1/f = 1/250 = 0.004$ second or four milliseconds.

Ninety degrees is a quarter of a cycle, so it corresponds to a time of one millisecond. Thus, saying that *one of these* signals lags the other by 90 degrees is exactly the same as saying that it occurs one millisecond later. Note that the second signal in Fig. 4-6 reaches its peak one millisecond later than the first signal.

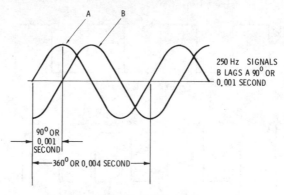

Fig. 4-6. At 250 Hz 90° is 0.001 second.

Now Fig. 4-7 shows exactly the same situation as Fig. 4-6, but the time scale is different. This time the two signals have a frequency of 1000 Hz. Now the duration of each complete cycle is one quarter of what it was with the 250-Hz signals. The duration of a complete cycle is one millisecond, and the time delay corresponding to 90 degrees is only one-quarter of a millisecond.

Two facts are apparent from these two examples:

1. Saying that two signals differ in phase is the same as saying that one of them is delayed in time with respect to the other.

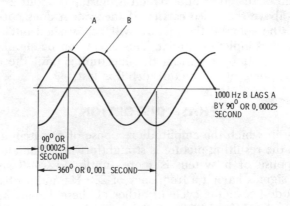

Fig. 4-7. At 1000 Hz 90° is 0.00025 second.

2. The amount of time involved in a given phase shift depends on the frequency. At 250 Hz, 90 degrees means one millisecond, whereas at 1000 Hz, 90 degrees means one quarter of a millisecond.

No real physical system will pass a signal instantly; some time is required, even though it is a very short time. Thus, all signals will be delayed when passing through a cable tv system. In order to preserve the shape of a waveform as it passes through a system, all components of the signal must be delayed by the same amount of time. Since a given amount of phase shift in degrees represents a different time delay at different frequencies, a system that delays all signals by the same amount of time will have a phase-versus-frequency curve that is a straight line, like the one shown in Fig. 4-8.

At first it might appear than any electrical system would be inclined to delay all frequencies by the same amount. If the system were purely resistive, this would indeed be true. However, when a system contains inductance and capacitance, as all cable systems do, the situation is different. Some frequencies will be delayed more than others. Fig. 4-9 shows both the amplitude and phase response of a low-pass filter of the type that might be used in a two-way cable tv system to separate upstream and downstream signals. Note that the phase-versus-frequency curve (Fig. 4-9B) is linear over part of the frequency range, but varies sharply in the range where the amplitude response changes rapidly.

Earlier we showed that a pulse type of signal might be thought of as consisting of a fundamental sinusoidal signal to which various amounts of its harmonics had been added. What we did not mention is that in order to produce the pulse shape, the harmonics that are added must have the proper phase. If the phase of a system does not increase linearly with frequency, some parts of the pulse will be delayed more than others, and the shape of the pulse will be distorted. Fig. 4-10 shows the way in which a nonlinear phase characteristic will distort a pulse.

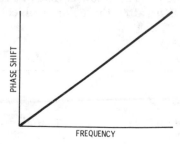

Fig. 4-8. Ideal phase-versus-frequency response of a tv system.

The effect of phase distortion on color signals is more involved. Each portion of a television signal can be thought of as consisting of two different parts. First, there will be the luminance that tells how bright that part of the scene is. In addition, there will be the chroma information modulated onto the 3.58-MHz subcarrier. This part of the signal tells the hue and saturation of that portion of the scene. It is obvious that as each portion of a scene is displayed on the screen, both the luminance and color information must be displayed at the same time. Any delay in the chrominance signal with respect to the luminance signal will result in improper color registration. The color will tend to lag the object on the screen.

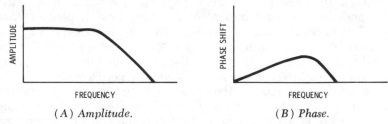

(A) Amplitude.　　　　　　　　(B) Phase.

Fig. 4-9. Response of typical low-pass filter.

The way this effect takes place can be seen in Fig. 4-11. Here we have a nonlinear phase characteristic of some part of a system. For the sake of illustration, suppose that the brightness of a part of a scene is changing at such a rate that the luminance signal will have a 1-kHz component. Naturally the chrominance will also have a 1-kHz component, but it will be transmitted through the system modulated on the 3.58-MHz color subcarrier. The phase shift that the color information will experience is the difference between the phase shifts at 3.58 MHz and at a frequency 1 kHz away.

Referring to Fig. 4-11, the luminance component at 1 kHz above the bottom of the band will be delayed by 0.2 μs. The chrominance information, being transmitted at approximately 3.58 MHz, will be delayed by nearly 0.4 μs. The result is that the color will appear on the screen about 0.2 μs after the colored object. This amounts to

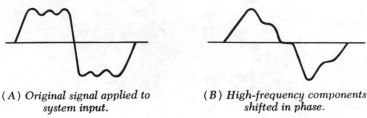

(A) Original signal applied to　　　(B) High-frequency components
system input.　　　　　　　　　　shifted in phase.

Fig. 4-10. Effect of phase distortion on a pulse signal.

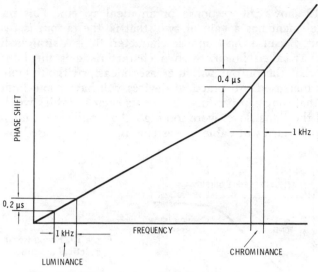

Fig. 4-11. Effect of phase distortion on color signals.

0.3% of a horizontal line, or about 1/10 inch on a 25-inch picture tube.

A term that is widely used, but often misunderstood, in cable tv is *envelope delay*. This is merely the amount by which the entire signal is delayed while passing through the system. Obviously, the envelope delay should be constant for minimum distortion of the television picture.

LINEARITY

All of the above discussion assumed that the response or output of any part of the system was proportional to the input. In actual practice, this can only be true over a limited range of operation.

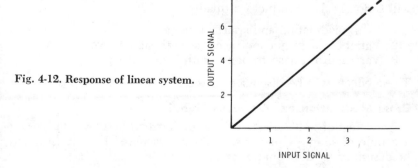

Fig. 4-12. Response of linear system.

Fig. 4-12 shows the response of an ideal system. This particular system element has a gain of two; that is, the output is two times the input. Because the response characteristic is a straight line, the response is called *linear*. Such a characteristic is ideal in that it implies that the output will increase linearly without limit as the input is increased. All practical devices will have some limit to the output that they can produce. Elements such as cables would break down if the signal level were too high. An amplifier would saturate. Most nonlinearity in cable tv systems is, in fact, caused by amplifiers.

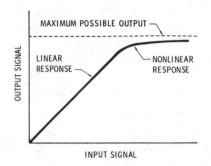

Fig. 4-13. Response of a
practical amplifier.

Fig. 4-13 shows the response characteristic of a practical system element. Note that the curve flattens out after the input reaches a certain level. This means that the device, which might be an amplifier, can produce only a certain output, regardless of how much input is applied. The implication of this is that amplifiers, converters, etc., must be operated over a range where the response characteristic is the most linear. As we will see in the chapter on amplifiers, no amplifier is perfectly linear over any part of its range. There is always some nonlinearity, however small. In a cable tv system, many amplifiers are cascaded, and any trouble due to nonlinearity in the individual amplifiers will be compounded.

There are three effects of nonlinearity in a cable tv system that will seriously degrade picture quality:

1. Cross modulation and spurious signal
2. Variation of frequency response with signal level
3. Variation of phase response with signal level

These effects will be discussed in the following paragraphs.

Cross Modulation and Spurious Signals

The effect of nonlinearity can be understood best by considering a couple of simple equations. A linear response characteristic can be described by the equation

$$O = KI$$

where,

O is the output,
I is the input,
K is the gain or loss of the system.

Any nonlinear response characteristic, no matter how curved it may be, can be described by an equation of the form:

$$O = K_1I + K_2I^2 + K_3I^3 + \text{more terms}$$

The first term is the same as the response of a linear system. The additional terms describe how much the characteristic differs from a straight line. Thus, the terms K_2, K_3, etc., tell the kind and amount of nonlinearity and should be kept as low as possible. This is accomplished both by design and by operation of the cascaded amplifiers over a range more limited than the one used with a single amplifier.

If an amplifier has a response such that the right-hand side of the above equation has only two terms, and the value of K_2 is a substantial fraction of K_1, an effect known as second order distortion will occur. If only a single signal is applied to the amplifier, the output will contain not only an amplified version of the input signal but also a second-harmonic component. In a cable tv system that uses all channels, this second harmonic might interfere with another signal. If two signals are applied, the situation will be complicated considerably. In addition to the two input signals and their second harmonics, we will obtain signals having frequencies equal to the sum and difference of the input frequencies. Since more than two signals are involved, the situation quickly becomes unmanageable.

When the characteristic is such that the equation has more than two terms on the right-hand side, still more spurious frequencies are generated, but in this case another phenomenon, known as *cross modulation,* occurs. With second order distortion, spurious signals were generated, but the original input signals were present in the output in substantially the same form they entered the system. With cross modulation, the modulation of one signal also appears on the other.

The requirements for linearity will be discussed in more detail in the chapter on cable amplifiers.

Incremental and Differential Gain

Fig. 4-14 shows an amplifier characteristic together with the input and output waveforms. In this case, the input signal level is high enough to drive the amplifier into its nonlinear region. Note how the sync pulses tend to become compressed. If the level of the input signal were increased, both the luminance and chrominance portions of the signal would be compressed. In the extreme case, both color and signal detail would be lost, if the signal could be maintained in synchronism at all.

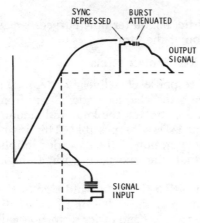

Fig. 4-14. Effect of incremental gain.

Two terms are used to describe this effect. *Incremental gain* refers to the change in the slope of the characteristic of a device. If the system is linear, the slope is constant and the incremental gain is zero. *Differential gain* refers to the same thing, but it is expressed in terms that permit easy measurement and can be related to the resulting distortion. The differential gain of a system is the difference in gain of a small high-frequency signal—usually the color subcarrier—when it is superimposed on two different levels of a low-frequency signal—usually the horizontal line frequency. Ideally, the differential gain of a system would be zero.

Incremental Phase Distortion

It was pointed out earlier that ideally the phase-shift versus frequency characteristic of a cable tv system should be a straight line. In a system with nonlinearity, signal level can affect the amount of phase shift, and hence delay, that a signal experiences. This means that high-level signals may be delayed more or less than low-level signals. In other words, in a nonlinear system, bright portions of an object may be displaced from darker portions. This is called *incremental phase distortion*.

This effect is specified and measured in terms of *differential phase*. It is defined as the difference in phase of a small high-frequency signal—usually the 3.58-MHz color subcarrier—when superimposed on two different levels of a low-frequency signal—usually the horizontal line frequency. It can be seen that one effect of differential phase is to make the hue of an object depend on the brightness. Ideally, differential phase should be zero.

NOISE

One of the most important characteristics of a cable tv system is the amount of noise it introduces into a signal. This consideration is intimately related to amplifiers and their spacing and, for this reason, is discussed in the chapter on cable tv amplifiers.

CHAPTER 5

Coaxial Cable
Transmission Lines

The very heart of any cable tv system is the coaxial cable transmission line. All of the other components of the system are used either to get signals into or out of the cable or to overcome some of the basic limitations of the cable itself. A firm understanding of coaxial transmission lines is thus essential to an understanding of cable tv.

Coaxial cables are not new. They have been used for many years in telephone systems and in radio and television broadcasting. However, in most of these applications the length of the cable has been short or the bandwidth of the signals small, or both. In the cable tv application, the bandwidth is very large and the cables are comparatively long. These conditions, together with the fact that the requirements for faithful transmission of a television signal are stringent, pose some rather unique problems that are not encountered in other cable applications.

BASIC PRINCIPLES

The purpose of a transmission line is to carry electrical energy from one point to another with a minimum amount of loss. At low frequencies, the job is easy, and just about any arrangement of conductors of the proper size will do the job. As the frequency becomes higher, new difficulties arise. At radio frequencies, any wire longer than about one-tenth of a wavelength will try to act as an antenna. That is, it will both radiate and receive signals.

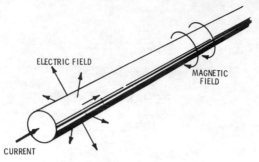

ELECTRIC FIELD

MAGNETIC
FIELD

CURRENT

Fig. 5-1. The fields around a conductor.

Naturally, this is undesirable in a cable tv system because radiation would reduce the signal level in the system and interfere with other services, and reception would interfere with the television signals being carried.

Fig. 5-1 shows the fields around a current-carrying conductor. Electric lines extend radially from the conductor, and magnetic lines surround it. In a conductor longer than one-tenth wavelength, these fields acting together cause radiation. The earliest form of transmission line that was widely used to minimize radiation was the two-wire line shown in Fig. 5-2. Although this type of line was never used in cable tv, it will serve to illustrate an important aspect of transmission lines of all types. In the two-wire line, the currents at any point along the line are equal and in opposite directions. As a result, the electric and magnetic lines surrounding the two conductors are opposite to each other, as shown in Fig. 5-3, and tend to cancel each other.

If an observer were located some distance away from a two-wire line, he would be unable to distinguish two separate wires. He would notice just a single wire. At this distance, the two opposing electric and magnetic fields could no longer be distinguished, but would, for all practical purposes, be completely cancelled by each other. As the observer approached closer to the line, he would begin to notice that there were actually two wires, rather than just one. At this distance, measurements would indicate that the fields

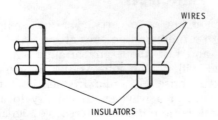

WIRES

Fig. 5-2. A simple two-wire
transmission line.

INSULATORS

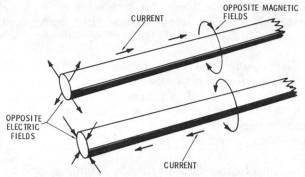

Fig. 5-3. Currents and fields in a two-wire line.

from the two wires did not completely cancel and a resultant field could be measured.

Fig. 5-4 shows the basic construction of coaxial cable. Like the two-wire line, it consists of two conductors, but here one conductor is located inside the other. The inner conductor is a wire and can be either solid or stranded. The outer conductor completely surrounds the inner conductor like a pipe or conduit. It can be either solid or braided. It is usually covered with a plastic material or jacket to protect it against the weather. The inner conductor is held in

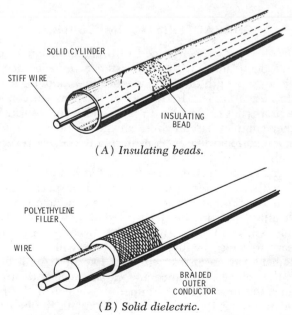

(A) Insulating beads.

(B) Solid dielectric.

Fig. 5-4. Two typical coaxial cables.

position at the center of the outer conductor either by insulating beads spaced periodically along the line (Fig. 5-4A) or by a solid dielectric material that fills the space between the conductors (Fig. 5-4B).

As in the two-wire line, the currents in the two conductors are equal in magnitude and opposite in direction at every point along the line. Here again, the fields from the two conductors tend to cancel. A detailed mathematical analysis would show that this cancellation occurs at the surface of the outer conductor. Thus, all of the field is contained within the cable, and there is no radiation or pickup.

This concept of cancellation of fields is fundamental to all types of electronic shielding. The signal is kept where it belongs by equal and opposite fields that cancel each other. It is *not* accomplished in the way that a man is protected from the rain by a raincoat. This approach makes it easy to understand why any disturbance in a system that results in unequal currents in the two conductors of a cable will result in radiation from the cable. It is not immediately obvious that an arrangement that will minimize radiation from a system will also minimize pickup of signals from outside the system, but it will. It is a good rule of thumb that a cable system that radiates badly will also be subject to interference from strong signals outside the cable.

CHARACTERISTIC IMPEDANCE

In dc circuits we are accustomed to thinking of connecting wires and lines as being ideal with no resistance or reactance. With ac signals, when the length of a line is more than about one-tenth of a wavelength at the highest frequency of interest, we can no longer ignore the properties of the line. In fact, such a line looks electrically like a rather complex circuit. We can get some idea of what this equivalent circuit looks like by considering the physical properties of the line itself.

Each of the conductors in a coaxial cable has appreciable length compared with a wavelength, so we will suspect that it has a significant amount of inductance. Since the conductors are insulated from each other, but are in close proximity, we will also suspect that the line will have some capacitance. It is not surprising, therefore, to learn that electrically a coaxial line, which we will assume for the moment has no losses, will look like the circuit of Fig. 5-5. Of course, the actual inductance and capacitance is distributed uniformly along the line and not lumped as shown in the figure, but the equivalent circuit is good enough to explain the behavior of an actual line.

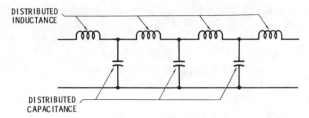

Fig. 5-5. The equivalent circuit of a coaxial line.

An easy way to see how the inductance and capacitance of a coaxial line influence the electrical behavior of the line is to consider a fictitious line that is infinitely long. Of course, no actual line can be infinitely long, but the concept will make several properties of the line clear. Fig. 5-6 shows a section of an infinite line connected to a voltage source. The line itself looks like an infinite number of capacitors connected through an infinite number of inductors. There are two properties of capacitances and inductances that are useful in studying transmission lines:

1. The voltage across a capacitor cannot change instantly; it depends on a charge being built up. Thus, if a capacitor is suddenly switched into a circuit, it will look like a short circuit at the first instant until it has had time to take on a charge.

2. The current in an inductance cannot change instantly. Any change in current is opposed by the counter electromotive force (emf) produced by the magnetic field associated with the inductance. Thus, if an inductance is suddenly switched into

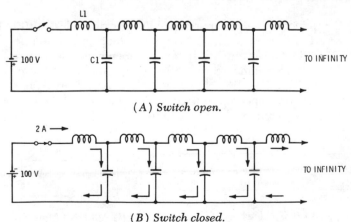

(A) Switch open.

(B) Switch closed.

Fig. 5-6. Current in an infinite line.

a circuit, it will look like an open circuit the first instant after the switch is closed.

Now, getting back to our infinitely long line in Fig. 5-6, we see that at the instant the switch is closed, capacitor C1 looks like a short circuit and inductance L1 looks like an open circuit. Current will start to charge C1, but the current will not be infinitely large because it is opposed by inductance L1. The process will continue indefinitely because there are an infinite number of capacitors to be charged. There will be a definite relationship between the applied voltage and the amount of current that will flow. The relationship will depend only on the values of L and C, which in turn depend on the physical dimensions of the line. In our example of Fig. 5-6, an applied voltage of 100 volts will cause a current of 2 amperes to flow into the line when the switch is closed. As far as the source is concerned, it has no way of knowing whether it is connected to a transmission line that is infinitely long or to a 50-ohm resistor as shown in Fig. 5-7. In both cases, a current of 2 amperes would flow. For this reason, we say that this particular line has a *characteristic impedance* or a *surge impedance* of 50 ohms.

The characteristic impedance depends only on the values of L and C in the equivalent circuit, and these in turn depend on the dimensions of the line. The value is given by the equation:

$$Z_0 = \sqrt{L/C}$$

where,
 Z_0 is the characteristic impedance in ohms,
 L is the inductance,
 C is the capacitance.

Since the inductance and capacitance are distributed along the line, they are given in units such as henrys per foot and farads per foot. Since the inductance and capacitance depend on the construction of the line, the impedance can also be determined from the physical dimensions of the line. Fig. 5-8 gives a curve and an equation for the characteristic impedance of coaxial cables as a function of the ratio of the diameters of the two conductors.

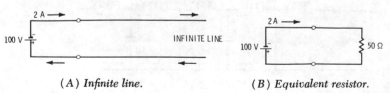

(A) *Infinite line.*   (B) *Equivalent resistor.*

Fig. 5-7. Characteristic impedance of an infinite line.

It is important to realize that the characteristic impedance of a transmission line is a function of the construction of the line itself; it does not depend on the signal carried or on what is connected to the line. So far, we know that if the line happens to be infinitely

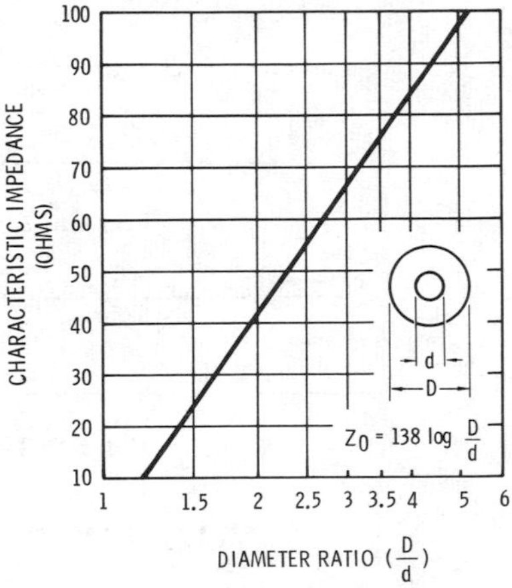

Fig. 5-8. Characteristic impedance of an air-filled line.

$$Z_0 = 138 \log \frac{D}{d}$$

DIAMETER RATIO $(\frac{D}{d})$

long, this is the impedance we will see looking into the line. We will now consider what we would see if the line were not infinitely long.

TERMINATION

All practical transmission lines must end somewhere. Our infinite line is strictly a fictitious device for understanding the principles. What we actually connect to the far end of a line is known as its *termination*.

Fig. 5-9A again shows our infinite line. If we were to cut the line at any point, say at point A, the section of the line to the right of where we cut it would still be infinitely long, so its impedance measured at this point would still be 50 ohms as shown in Fig. 5-9B. We now see that if we terminate the section to the left of our cut with a 50-ohm resistor, as shown in Fig. 5-9C, it will look like the infinite line of Fig. 5-9A as far as any measurements we can make at its input terminals are concerned. The line is then said to be *terminated in its characteristic impedance*. We now see that if a transmission line is terminated with a resistance equal to its characteristic impedance, its input impedance will be equal to its characteristic impedance. When the termination has any other value of impedance, the input impedance will have some other value, as we will see later.

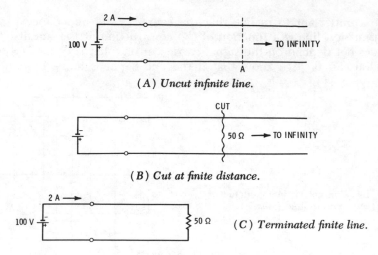

(A) Uncut infinite line.

(B) Cut at finite distance.

(C) Terminated finite line.

Fig. 5-9. Terminated line.

VELOCITY OF PROPAGATION

When a signal is propagated through a transmission line, the entire process of charging capacitances takes some finite time. We cannot think of the signal as traveling through the line at infinite velocity. Radio waves traveling through space have a velocity of about 186,000 miles, or 300,000,000 meters, per second. An equation from physics tells us that this figure is given by

$$c = \frac{1}{\sqrt{\mu\epsilon}}$$

where,
   c is the velocity of propagation,
   μ is the magnetic permeability of free space,
   ϵ is the dielectric constant of free space.

From this equation, we can guess that if there were no insulation between the conductors of a coaxial line, only free space, the velocity of propagation in the line would be the same as in free space. We cannot build a line, however, without some form of insulation to hold the inner conductor in position. Thus, even if we use small beads spaced periodically along the line, the average dielectric constant will be slightly higher than that of free space. Since this appears in the denominator of our equation, we would suspect that the velocity in the cable would be slightly lower than in free space. This is indeed true, and such a line has a velocity of propagation of about 97% of that in free space.

By the same reasoning, we would realize that if the space between the two conductors were filled with a dielectric material, the dielectric constant would be much greater than that of free space, and the velocity would be considerably lower. This is true. In solid coaxial lines, the velocity of propagation is about 60% of that in free space.

REFLECTIONS

One transmission-line consideration that is commonly misunderstood is the phenomenon of reflections. Fig. 5-10A shows a transmission line connected to a source at one end and terminated in its characteristic impedance at the other. We know that when the switch is closed, energy will enter the line. In this case, since the voltage is 100 volts and the current is 2 amperes, we know that the power will be 200 watts. Because of the velocity of propagation, we know that some time after the switch is closed, the energy will be dissipated in the terminating resistor. Since we have assumed that our ideal line has no losses, the question arises as to what happens to the power between the time the switch is closed and the time it reaches the terminating resistor. The answer is that during this period the energy is *stored in the line.*

Suppose now that we have the situation shown in Fig. 5-10B, where there is no terminating resistance. At the moment the switch is closed, the source cannot tell what is connected to the line. Power will flow into the line at the same rate as before. In this case, however, there is no place for the power to be dissipated. We have assumed that the line itself has no losses. Energy has definitely entered the line, but there is no place for it to be dissipated. Where does it go? The answer is that it is reflected back to the source.

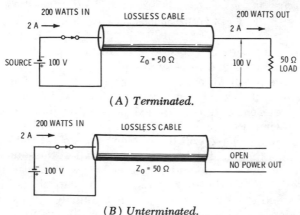

(A) *Terminated.*

(B) *Unterminated.*

Fig. 5-10. A comparison of a terminated and an unterminated line.

73

The way in which the termination of a line will affect the reflection of power can be understood by first considering how energy is stored in the line. Fig. 5-11A shows a section of a transmission line. This could be a section of an infinite line or a line terminated in its characteristic impedance. Notice that there is a constant current of 2 amperes through the inductances and a voltage of 100 volts across the capacitors. Under these conditions there is energy stored in the magnetic field around the inductances and in the electric field in the capacitors. Now suppose that this section is the final section of a line and that there is no termination at all. At the instant the signal reaches the end of the line, the situation will be as shown in Fig. 5-11A, but after capacitor C1 is charged to the full 100 volts, there is no place for the current to flow. The current flowing in L1 will therefore drop suddenly to zero. This will induce a counter emf in the line that will add to the voltage in C1 and will double it as shown in Fig. 5-11B. The current in the preceding inductance will then drop to zero, and the process will be repeated. Thus, the reflected wave will travel back along the line in a manner analogous to a string of dominoes toppling each other when one of them is tipped over. A similar reflection will be initiated when the end of the line is shorted.

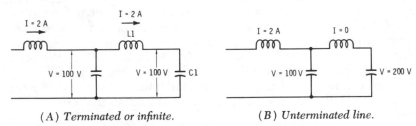

(A) Terminated or infinite.   (B) Unterminated line.

Fig. 5-11. Reflection in an open line.

A quantitative idea of reflections and how they depend on the terminating resistance can be gathered from Fig. 5-12. Here the source is more practical in that it has an internal impedance, which all real sources do. In this case both the internal impedance of the source and the characteristic impedance of the line are 50 ohms. When the switch is closed (Fig. 5-12A), current will flow into the line, and a voltage wave will travel along the line at the velocity of propagation of the line. During this period, half of the source voltage will appear across the line and the other half will appear across the internal resistance of the source. Each of the capacitors in the line will be charged to 50 volts. If the line is terminated in its characteristic impedance, the final state will be as shown in Fig. 5-12B. Each of the capacitors in the equivalent network will be charged to 50 volts; there will be 1 ampere flowing in the inductances, and there will be no reflection.

Now consider the situation of Fig. 5-12C, where the far end of the line is open. The situation will be the same as in Fig. 5-12A until the voltage wave reaches the end of the line. At this time the field associated with the last inductance in the equivalent circuit will collapse, inducing an additional voltage and charging the line to 100 volts (Fig. 5-12D). After the reflection reaches the source, the final state of the line will be as shown in Fig. 5-12E. The line will be charged to 100 volts, there will be no current, and hence there will be no voltage drop across the internal resistance of the source.

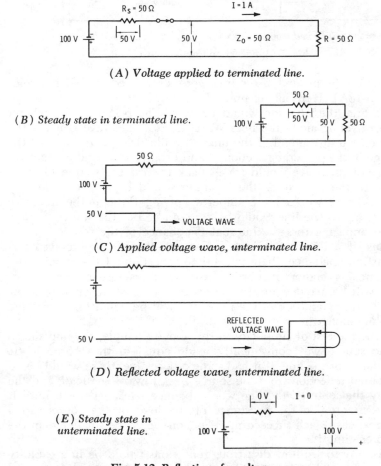

(A) Voltage applied to terminated line.

(B) Steady state in terminated line.

(C) Applied voltage wave, unterminated line.

(D) Reflected voltage wave, unterminated line.

(E) Steady state in unterminated line.

Fig. 5-12. Reflection of a voltage wave.

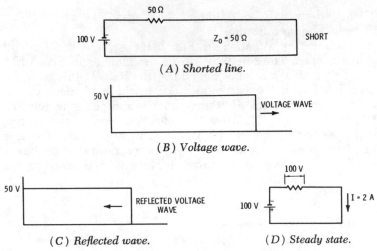

(A) Shorted line.

(B) Voltage wave.

(C) Reflected wave.　　　　　　　(D) Steady state.

Fig. 5-13. Reflection from a short circuit.

Suppose the far end of the line were terminated in a short circuit (Fig. 5-13A). Here again, until the voltage wave reached the end of the line, the situation would be as in Fig. 5-13B with a 50-volt wave traveling along the line. When the voltage wave reached the short circuit at the end of the line, it would drop to zero. Next, the voltage in the preceding section would drop to zero, and a negative-going voltage wave would travel back toward the source (Fig. 5-13C). In the final state, the situation would be as shown in Fig. 5-13D. There would be 2 amperes flowing through the line, the voltage across the line would be zero, and all of the source voltage would appear across the internal impedance of the source.

Thus, if a line is terminated in a resistance equal to its characteristic impedance, there will be no reflection. If the terminating resistance is higher or lower than the characteristic impedance, there will be a reflection. In the extreme case of an open or short circuit, the magnitude of the reflection will be equal to the voltage traveling along the line.

In the interest of simplicity in the above example, the impedance of the source was conveniently made equal to the characteristic impedance of the line. If it had some other value, there would be an additional reflection from the source back toward the load, and the energy that entered the line would bounce back and forth until it was dissipated in the resistance of the load or the source. In a practical case, each succeeding reflection would be smaller than the one preceding it.

It is easy to see how disastrous reflections would be in a cable tv system. The television signals would bounce back and forth and arrive at the receiver at different times, causing ghosts that would be very objectionable. There are two ways of minimizing the reflections. The first is to be sure that all lines are terminated in their characteristic impedance. In this way there will be no reflection. The other way is to match the internal impedance of the source to the line. This will not stop reflections, but it will ensure that they will be dissipated when they get back to the source and will not return to the line. In practice, both techniques are used.

LOSS AND ATTENUATION

The preceding discussion of transmission lines completely neglected any losses that might occur in the line itself. Not only do practical cables have loss, but the nature of the loss is such that it has serious implications in cable tv. Fig. 5-14 shows a sketch of a coaxial cable with the two components of loss shown. The first and least important from our point of view is the loss in the dielectric material between the two conductors. This loss is unimportant because it is small and usually does not need to be considered except

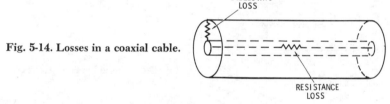

Fig. 5-14. Losses in a coaxial cable.

at the very highest frequencies encountered in cable tv. The second cause of loss is the resistance of the conductors. Since the inner conductor is smaller, its resistance will be higher and it will contribute most of the loss.

The loss of a conductor at radio frequencies is complicated by a phenomenon known as *skin effect*. When a direct current flows through a conductor, the current is distributed uniformly throughout the cross section of the conductor, as shown in Fig. 5-15A. As the frequency becomes higher, the current tends to crowd toward the surface, as shown in Fig. 5-15B. This reduces the cross section through which the current actually flows, resulting in a given conductor having a higher resistance at higher frequencies. From this we can conclude that the loss in a coaxial cable will not be the same at all frequencies but will increase with frequency. An extensive mathematical investigation would show that the increase in resistance varies as the square root of the frequency. Fig. 5-16

(A) *Direct current.*

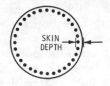
(B) *Alternating current.*

Fig. 5-15. Skin effect.

shows the loss of various types of coaxial cables as a function of frequency. The information is plotted on log-log paper where the curves conveniently become straight lines.

In considering the loss of a coaxial cable, it might appear that for a given size of outer conductor, we could reduce the losses by

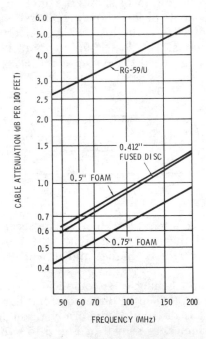

Fig. 5-16. Cable losses vs frequency.

making the inner conductor larger so that its resistance would be lower. To some extent this is true. However, as we increase the diameter of the inner conductor, the characteristic impedance of the cable becomes lower. This means that a higher current will be required to transmit the same amount of power. Since the loss is proportional to the square of the current, we will quickly reach a point where the loss is actually increased by increasing the diameter of the inner conductor.

If we were to take the opposite approach of reducing the diameter of the inner conductor in an attempt to increase the impedance so that the current would become smaller for the same amount of power, we would reach a point where the conductor resistance increased fast enough to increase the losses. These two conflicting considerations indicate that there is an optimum ratio of conductor sizes that will result in minimum loss. This is true, and with air-filled coaxial cable, the optimum ratio is about 3 to 1, which corresponds to a characteristic impedance of about 70 ohms. This is one of the reasons why cable tv systems almost universally use 75-ohm coaxial cable.

There are other factors that enter into the selection of an optimum characteristic impedance for a coaxial cable. One of them is the voltage rating. If the inner conductor is made too large, the gap between the conductors will be so small that an arc can form between them, or if the inner conductor is made too small, corona discharge will occur at its surface.

INPUT IMPEDANCE

We have already seen that if a transmission line is terminated in its characteristic impedance, this will be the impedance that we see looking into the line. This is true regardless of the length of the line. However, when a line is not terminated in its characteristic impedance, the input impedance will have some other value. Actually, this impedance can be almost any value because it will depend on the length of the unmatched line in wavelengths.

This variation in impedance can be appreciated by the fact that in an unmatched line there are reflections. With a sinusoidal ac signal, the reflection will also be sinusoidal and will add to or subtract from the applied signal. Since the phase of the reflected signal depends on the length in wavelengths, it is easy to see that the input impedance will be a function of frequency.

Since a cable tv system carries signals having a wide frequency range, it is imperative that the lines be as well matched as possible at both ends.

VSWR, REFLECTION FACTOR, AND RETURN LOSS

There are several different ways to express the effect of an impedance mismatch on the performance of a transmission line. They all say the same thing in different ways. A measurement commonly used in communications systems is the *vswr*, or *voltage standing wave ratio*. If a transmission line is connected to an ac source as shown in Fig. 5-17, and the impedance connected to the far end

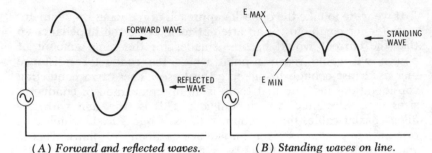

(A) Forward and reflected waves.   (B) Standing waves on line.

Fig. 5-17. Formation of standing waves.

does not match the characteristic impedance of the line, reflections will occur. The effect is that of a sinusoidal signal traveling down the line from the source, and another sinusoidal signal traveling from the load back toward the source. The two signals add algebraically all along the line with the result that there is a standing wave set up on the line.

The way in which the standing wave is generated can be easily understood by means of an analogy with waves of motion along a rope, as shown in Fig. 5-18. Suppose that the far end of a rope is not attached to anything and the rope is given a shake. A single wave of motion will travel along the rope as shown in Fig. 5-18A. If the shaking is continued rhythmically, a traveling wave will be sent down the rope as in Fig. 5-18B. Suppose that the far end of the rope is securely fastened and the rope is given a single shake

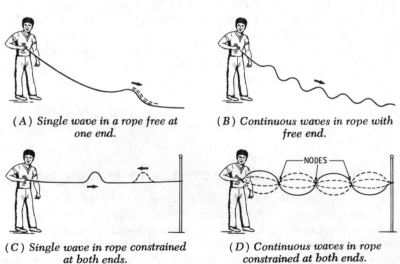

(A) Single wave in a rope free at   (B) Continuous waves in rope with
        one end.   free end.

(C) Single wave in rope constrained   (D) Continuous waves in rope
        at both ends.   constrained at both ends.

Fig. 5-18. Wave motions in a rope.

as in Fig. 5-18C. A single wave will travel down to the end of the rope as shown and will be reflected back toward the starting end. If the rope is shaken rhythmically, as in Fig. 5-18D, its motion will be the sum of the motion due to the direct wave and the motion due to the reflected wave. Since the two waves are traveling at the same speed, one traveling toward the far end and one coming back, the net result is that the wave does not move at all. There is a standing wave on the rope. The rope vibrates between nodes that do not vibrate at all.

At every point along the rope, the actual displacement is due to the combined effects of the direct and reflected waves. The actual displacement at any point can be found by algebraically adding the displacement due to the direct wave acting alone and the displacement due to the reflected wave alone. The nodes occur when the algebraic sum of the two waves is zero. Voltage and current waves on a transmission line combine in the same way to produce standing waves.

The term vswr is simply the ratio of the maximum voltage along a line to the minimum value. It is given by:

$$\text{vswr} = \frac{E_{max}}{E_{min}}$$

Another term that is also used to express the effect of mismatch is the *reflection coefficient*, ρ. This is the ratio of the reflected voltage to the incident voltage (that is, the forward wave of voltage) at the load. It is given by

$$\rho = \frac{E_r}{E_i}$$

where,

E_r is the reflected voltage,
E_i is the incident (forward) voltage.

A term that is used frequently in cable tv is *return loss*, R. This term is misleading and should be properly understood. It is merely the value of the reflection coefficient expressed in decibels. It is given by

$$R = -20 \log \rho$$

or

$$R = 20 \log 1/\rho$$

When the reflection is minimal, the return loss will be numerically larger.

Vswr, reflection coefficient, and return loss are all just different ways of expressing the same phenomenon. The quantity used is

Chart 5-1. VSWR, Reflection Coefficient, and Return Loss

Standing-Wave Ratio (vswr)

$$\text{vswr} = \frac{E_{max}}{E_{min}} = \frac{(1 + \rho)}{(1 - \rho)}$$

$$= \frac{Z_L}{Z_0} \text{ for } Z_L > Z_0$$

$$= \frac{Z_0}{Z_L} \text{ for } Z_L < Z_0$$

Reflection Coefficient (ρ)

$$\rho = \frac{E_r}{E_i}$$

$$= \frac{(Z_L/Z_0 - 1)}{(Z_L/Z_0 + 1)} \text{ for } Z_L > Z_0$$

$$= \frac{(Z_0/Z_L - 1)}{(Z_0/Z_L + 1)} \text{ for } Z_L < Z_0$$

Return Loss (R)

$$R = -20 \log \rho$$

$$= 20 \log 1/\rho$$

selected for convenience in manipulation or correlation with actual measurements. The equations relating these quantities are given in Chart 5-1.

IMPEDANCE AND REFLECTIONS IN PRACTICAL CABLES

All of the foregoing discussion in this chapter was centered around more or less ideal cables. Loss was not considered, except in the section dealing with attenuation in cables.

The characteristic impedance of a coaxial cable without any losses is given by the equation:

$$Z_0 = \sqrt{L/C}$$

It is interesting to note that the characteristic impedance is resistive; that is, it has no reactive component. What this means is that when the line is properly terminated, all of the energy goes through the line in one direction. Or, to put it another way, all of the energy that is stored in the inductance and capacitance of the line eventually find its way into the load.

A more exact expression for the characteristic impedance of the line and one that takes losses into consideration is:

$$Z_0 = \sqrt{\frac{R + j\omega L}{G + j\omega C}}$$

Here, $R + j\omega L$ is the series impedance, and $G + j\omega C$ is the shunt admittance (Fig. 5-19), both per unit length. In practice, the series resistance in ohms and the shunt conductance, G, in mhos are both very small numbers, and our original equation was based on the assumption that $LG = RC$, which is very nearly true. It is worth noting, however, that when this simplifying assumption is not met, the characteristic impedance will be changed slightly and will have a reactive component. The presence of the reactance indicates that some of the energy that is stored in the equivalent inductance or capacitance of the line does not reach the load but is dissipated in the line.

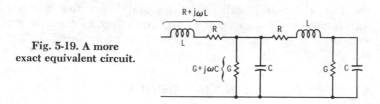

Fig. 5-19. A more exact equivalent circuit.

In a practical cable tv system, lines are terminated as closely as possible in their characteristic impedance. This is not always easy to accomplish in practice because of the wide range of frequencies used. By the same token, the dimensions of the line are held to close tolerances so that the inductance and capacitance per foot will be as constant as possible.

Any discontinuity in a cable will give rise to a reflection, even though it may be small. Fig. 5-20A shows a section of transmission line. The section to the right of the line A-A is assumed to be perfect and to be terminated in 50 ohms, the characteristic impedance of

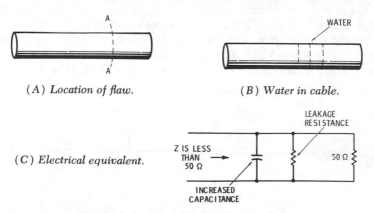

(A) Location of flaw.

(B) Water in cable.

(C) Electrical equivalent.

Fig. 5-20. The effect of a flaw in the cable.

the line. Suppose that there is some sort of discontinuity in the line at A-A. It might be that a lightning surge had punched a small hole in the outer conductor and that water, which has a dielectric constant of about 80, had leaked into a short section of the cable as shown in Fig. 5-20B. The section to the left of A-A can then be though of as a separate line that is terminated in two separate impedances. One is the 50 ohms that represents the line to the right of the trouble; the other is the increased capacitance and probably leakage resistance due to the water in the cable. The combination of the two terminating impedances will be less than the characteristic impedance of the line, and a reflection will be set up.

In addition to reflections that result from discreet discontinuities in the line, there are small reflections and losses that result from slight nonuniformities along the line. The effect of these imperfections is usually stated in terms of *structural return loss* and is given in decibels.

COAXIAL DEVICES

In addition to the cable itself, special devices are used to get signals into and out of cables with a minimum amount of mismatch. Many of these devices operate on the principle of the directional coupler.

The directional coupler (Fig. 5-21) is a mystifying device that has three coaxial terminals. There is a signal at the tap that is proportional to the direct signal traveling through the line from left to right; however, the coupler will ignore signals traveling in the opposite direction—from right to left. There is usually some predetermined amount of loss between the direct signal in the cable and that appearing at the tap.

There are several different arrangements that can be used to make a directional coupler. The technician will not be interested in what is inside the coupler, but he will be vitally interested in how it works. The directional coupler arrangement shown in Fig. 5-22 is quite easy to understand. It consists of a section of coaxial cable

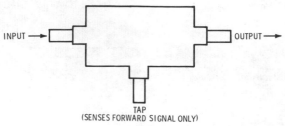

Fig. 5-21. Directional coupler.

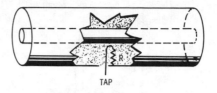

Fig. 5-22. The principle of a
directional coupler.

TAP

that has a probe inserted inside the outer conductor and terminated
in a resistor. If this resistor were infinitely large, we would think of
the probe as being one plate of a capacitor, and the inner conductor as being the other plate. This small capacitor would be in parallel
with the distributed capacitance of the line and would charge to the
voltage across the line. If, on the other hand, the resistance were
zero, the probe would look like a loop and could be thought of as
a one-turn secondary of a transformer. It would then respond only
to the current flowing in the line. In practice, the resistance is
adjusted so that the probe will respond equally to the voltage and
current in the line. As far as the direct wave is concerned, the
voltage and current are in phase, so their effects will add in the
probe. But in the reflected wave, the current is traveling in the
opposite direction—back toward the source. This means that the
effects of the voltage and current of the reflected wave in the probe
are opposite and tend to cancel. With proper adjustment, they will
cancel completely, and the coupler will be insensitive to reflected
signals. In practice, complete cancellation is not obtained over the
complete band of frequencies used in cable tv, but very high isolation can be obtained.

There are four parameters used to describe the performance of a
directional coupler: *insertion loss, tap loss, isolation,* and *directivity.*
These and the relationships between them are shown in Fig. 5-23.
They are usually expressed in decibels. In cable tv work, high isolation alone is not the important parameter. The directivity, which
is the difference between the isolation and the tap loss, is the
significant parameter. Many of the different coaxial devices that are
used to split or combine signals operate on the principle of the
directional coupler. These devices make it possible to connect various

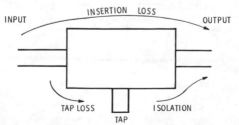

Fig. 5-23. The parameters of a directional coupler.

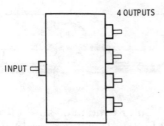

4 OUTPUTS

INPUT

Fig. 5-24. Four-way splitter.

system components together in a way that will provide isolation between them and keep impedances matched. Fig. 5-24 shows a four-way splitter. This device splits the signal equally into four different branches. Since the log of ¼ is −0.6, the signal in each of the outputs will be 6 dB below the signal into the device, if there is no loss in the device itself. In most cases there is a small insertion loss—about 1 dB.

Other devices of this type are used for various purposes in a complete system. They will be described in later parts of this book.

PRACTICAL CABLES

Many different types of coaxial cable are used in cable tv systems. In general, the selection of a particular type of cable represents a compromise between performance requirements and cost. In long

Courtesy General Cable Corp.

Fig. 5-25. Fused-disc coaxial cable.

runs, such as the trunk cables of a large system, the highest grade of cable is usually used. Any degradation of the signal will be aggravated by the length of the cable and will affect many subscribers. In short runs, such as drop cables to homes, however, the run is usually short, and less-expensive cable can be used.

Appendix B lists several types of cable commonly used in cable tv systems, together with their characteristics. Fig. 5-25 shows a new fused-disc cable that was developed specifically for cable tv.

CHAPTER 6

Cable TV Amplifiers

Amplifiers that are used in the distribution portion of a cable tv system have rather unique requirements. The required bandwidth is very large, the television signal is rather easily distorted, and usually several amplifiers are cascaded in the system. The fact that several amplifiers are cascaded means that the noise and distortion introduced by amplifiers early in the system are amplified by later amplifiers, which tends to aggravate the situation. These factors make the design and adjustment of cable tv amplifiers different from the design and adjustment of amplifiers used for other purposes.

The purpose of a cable tv amplifier is to compensate for the loss in a section of coaxial cable. Fig. 6-1 shows a portion of a cable tv system. The gain of each amplifier is numerically equal to the attenuation of the corresponding section of cable. This equality is carried throughout the system. If the gain of an amplifier were less than the loss of the cable, there would be a progressive loss of signal strength throughout the system, and a point would be reached where the signal-to-noise ratio would become low enough to seriously degrade the picture. If, on the other hand, the gain of an amplifier were greater than the loss of the corresponding section of cable, the signal strength would increase progressively along the system, and a point would be reached where the level would be such that it overloaded the amplifiers and introduced an unacceptable amount of distortion.

Since the amplifier gain in decibels is numerically equal to the loss of the section of cable between amplifiers, also in decibels, it is common to specify cable lengths in decibels instead of in feet. Thus, if a certain type of cable had a loss of 1 dB per 100 feet, a 1000-foot length of cable would be referred to as a 10-dB length. This practice of specifying cable length in decibels, while confusing at first, actually simplifies the understanding and maintenance of a system.

In specifying cable length in decibels, we neglected the important fact that cable attenuation is not the same at all frequencies. The attenuation varies with the square root of frequency and is always greatest at the highest frequency in use. Strictly speaking, we should also specify the frequency, and this is usually done in super-band systems. When a length of cable is specified in decibels without mention of frequency, the specification refers to the loss at the picture carrier frequency of channel 13.

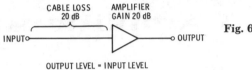

Fig. 6-1. Loss and gain in a
cable system.

The two parameters that have the greatest influence on cable tv amplifier design are the amount of noise the amplifier introduces and the amount that it distorts the signal. These factors determine what value of amplifier gain will result in the best system performance. To understand better how both noise signals and distortion products add up in a system, we will digress briefly and look at how different types of signals combine in a system.

COMBINING SIGNAL LEVELS

The simplest way to add the signals from two different sources is to connect two batteries in series as shown in Fig. 6-2. In Fig. 6-2A, a single 10-volt battery is connected to a simple 10-ohm load. The calculations in the figure show that the load current is 1 ampere and the power dissipated in the load is 10 watts. In Fig. 6-2B we have added another 10-volt battery in series with the first, with the polarity such that the two voltages will add. The voltage across the load is now 20 volts, the current is 2 amperes, and the power dissipated in the load is 40 watts. This follows from the fact that we doubled the voltage, and, since power is proportional to the square of the voltage, the power increased four times. In Fig. 6-2C, the second 10-volt battery is connected so that it opposes the first one, and the result is that the voltage across the load is zero and no power is dissipated.

There are really only two ways that we can connect two dc sources in series—either aiding or opposing. In one case, we get four times the power that we would get from one source, and in the

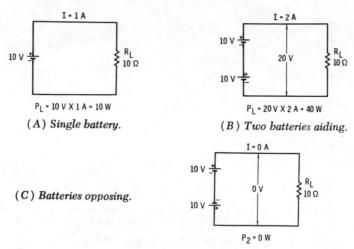

(A) Single battery.

$P_L = 10 V \times 1 A = 10 W$

(B) Two batteries aiding.

$P_L = 20 V \times 2 A = 40 W$

(C) Batteries opposing.

$P_2 = 0 W$

Fig. 6-2. Voltage and power in a dc circuit.

other we get nothing. With ac sources, the situation is different. There are an infinite number of ways the signals can add, depending on the phase angle between the two voltages.

In Fig. 6-3, two 10-V rms ac sources are connected in series across a 10-ohm load. If the two voltages are exactly in phase, the situation will be exactly the same as in Fig. 6-2B, and the power in the load will be 40 watts. Similarly, if the voltages are exactly out of phase, they will cancel each other, and no power will be dissipated in the load. If the phase angle between the two voltages is something between 0 and 180 degrees, the power dissipated in the load will be some value between these two extremes. The amount of power dissipated in the load for various phase angles is shown in Fig. 6-4. A very interesting point on this curve is the place where the phase angle between the two voltages is 90 degrees. Here, the power dissipated in the load is 20 watts, just twice what it would be if one of the sources were acting alone.

Mathematically, two sinusoidal signals that differ in phase by 90 degrees are said to be *uncorrelated*. With two uncorrelated signals, the power dissipated in a load is simply the sum of the powers each one would contribute if it were acting alone. As was shown, this

Fig. 6-3. Load power depends on phase angle between ac voltages.

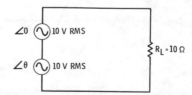

89

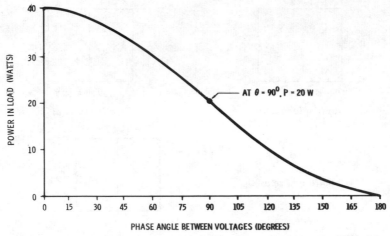

Fig. 6-4. Relationship between phase angle and power in circuit of Fig. 6-3.

does not work with other types of signals. Sinusoidal signals are of little interest to us at this point, but we are very interested in the way that uncorrelated signals combine.

Noise signals are of a random nature. As a result, noise that is caused by different devices is said to be uncorrelated. This means that noise signals from different amplifiers will combine in the same way that our uncorrelated ac signals combined. The powers can be added directly. The same thing is true of distortion products. We can add them on a power rather than a voltage basis. With this knowledge, we can see how noise and distortion products add up in a system that has several amplifiers cascaded together.

NOISE IN CASCADED AMPLIFIERS

Fig. 6-5 shows a section of a cable tv system. Each amplifier, no matter how well it is designed, will introduce some noise. For convenience, we can refer all of the noise to the input of the amplifier. To do this, we will make use of the concept of *equivalent noise input level*. This is the amount of noise that would have to be

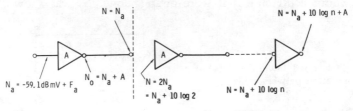

Fig. 6-5. Noise in cascaded amplifiers.

applied to the input of a perfect, noiseless amplifier to get the same level of noise at the output as we do from an actual amplifier. This noise level may be found by simply adding the thermal noise for a 4-MHz tv channel (-59.1 dBmV) to the noise figure of the amplifier in decibels:

$$N_a = -59.1 \text{ dBmV} + F_a$$

where,

N_a is the equivalent noise input level in dBmV,
F_a is the noise figure of the amplifier in decibels.

At the output of the amplifier, we have the sum of the equivalent noise input level and the gain of the amplifier in decibels:

$$N_o = N_a + A$$

The attenuation of the section of the cable following the amplifier is numerically equal to the gain of the amplifier, so at the end of the cable, the noise level will have dropped to the value it had at the input of the amplifier, that is, N_a.

Now the signal is applied to the input of the next amplifier, which also contributes an amount of noise equal to that contributed by the first amplifier—N_a. If we were to use the power notation, we could simply add the two powers because the two noise sources are uncorrelated. But, since we are working in dB and dBmV, which are logarithmic ratios of powers, addition is equivalent to multiplication; therefore, we will have to use a different technique which will be described in the following paragraphs.

When we are working with power, we can derive the level in dBmV by the following relationship:

$$N_a = 10 \log (P_n/P_r)$$

where,

P_n is the equivalent noise input power in watts,
P_r is the reference power used in establishing the dBmV—1/75 microwatt.

Now, if we added two noise powers, both equal to P_n, the level in dBmV would be given by:

$$N = 10 \log 2 \frac{P_n}{P_r}$$

A basic principle of mathematics tells us that multiplying two numbers is the same as adding their logarithms, so we can rewrite this equation as:

$$N = 10 \log (P_n/P_r) + 10 \log 2$$

or

$$N = N_a + 10 \log 2$$

This means the equivalent noise input level at the second amplifier is equal to that at the input of the first amplifier plus the quantity 10 log 2. As we might suspect, this principle is very general and can be applied to any number of amplifiers that are cascaded in such a way that the attenuation of the sections of cable between the amplifiers is equal to the gain of the amplifiers. Thus, in a system of n amplifiers, the noise level at the input of the nth amplifier is

$$N = N_a + 10 \log n$$

where N_a is the equivalent noise input level for each amplifier.

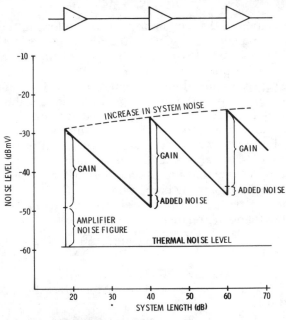

Fig. 6-6. Noise buildup in a system.

Note that this derivation was used to find the noise level at the input of any of the amplifiers in a system. To find the noise level at the output of the amplifier, simply add the gain A of the amplifier.

Fig. 6-6 shows a graph of how the noise level builds up in a system of cascaded amplifiers. In this illustration, the noise figure of each amplifier is assumed to be 10 dB, and each amplifier is assumed to have a gain of 20 dB. The bottom line in the figure is the unavoidable thermal noise. At the input to each amplifier, the noise is increased. As the noise passes through the cable, its level is decreased by the attenuation of the cable. At this point, the noise of the following amplifier is added, and the sum of the two is increased by the gain of the amplifier. The process then repeats with the noise level becoming progressively greater as we travel along the system. Although this type of plot shows the details of how the noise level behaves at each point in the system, it is more convenient to use straight lines to show how the noise level increases.

Fig. 6-7 shows a plot that gives the same information as Fig. 6-6. The bottom horizontal plot is the thermal noise. The next line is the noise level at the inputs of the amplifiers. The noise at the output of each amplifier will be equal to the noise at the input plus the gain of the amplifier in decibels. This is shown by the next line.

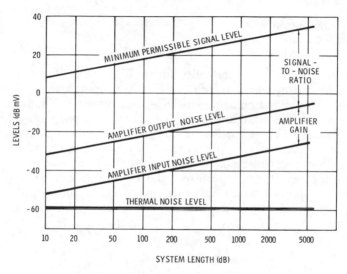

Fig. 6-7. Noise and signal levels.

Without justifying our statement at this time, we will say that to get a good-quality picture, the signal-to-noise ratio at all points in the system must be at least 40 dB. This means that the signal level at any point in the system must be greater than that of the sloping line in Fig. 6-7 labeled "Minimum Permissible Signal Level."

From Fig. 6-7, it can be seen that the longer the system, the higher the signal level needed to provide the required signal-to-noise level. Thus, noise considerations dictate how low the signal level may be and still provide a satisfactory picture. The maximum level that the signal may reach is dictated by the distortion characteristics of the amplifiers. These will be discussed in the following paragraphs.

LINEARITY AND DISTORTION

In an ideal amplifier, the output would at all times be directly proportional to the input. Whatever waveform was applied to the input would appear at the output with a larger magnitude, but with exactly the same shape. The output-versus-input characteristic of such an ideal amplifier would be a straight line as shown in Fig. 6-8A. Because the characteristic is a straight line, the amplifier is said to be linear. If we were to write an algebraic equation for the output voltage as a function of the input voltage, it would have the form

$$e_o = Ae_i$$

where,

e_o is the output voltage,
e_i is the input voltage,
A is the voltage gain of the amplifier.

No real amplifier is perfectly linear. There will always be some curvature to the characteristic. A typical amplifier might have an output-versus-input characteristic like that shown in Fig. 6-8B. This curve is nearly a straight line at low values of input voltage, but it curves more and more as the input voltage is increased. This means that at low values of input voltage, the waveform of the output voltage will be almost exactly the same as that of the input voltage. As the input voltage is increased, the waveform will become more distorted. Above a certain value of input voltage, the output voltage will not increase any further. At this point the amplifier is said to be *saturated*.

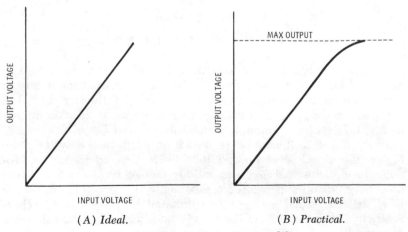

(A) *Ideal.*   (B) *Practical.*

Fig. 6-8. Characteristic of an amplifier.

To understand how a nonlinear amplifier characteristic will lead to distortion, it is helpful to look at the rather formidable equation

$$e_o = Ae_i + Be_i^2 + Ce_i^3 + \ldots$$

This equation is useful because it will describe any amplifier no matter how nonlinear it may be. To describe any curved line, all we have to do is add more terms to the right-hand side of the equation. The first term on the right side of the equation will be recognized as the output of an ideal amplifier. What our longer equation says, then, is that the output of a real amplifier consists of the output of an ideal amplifier *plus something else*. This something else is distortion.

The first type of distortion that we will examine is called *second-order distortion*. This means that only the first and second terms on the right side of our equation are large enough to be significant. The first term represents our desired output signal. The second term means that the output also contains other components. If only one signal is applied to the input, the spurious signal will be a

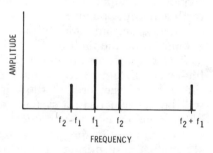

Fig. 6-9. Spurious signals resulting from second-order distortion.

second-harmonic component. If two signals are applied to the input, the output will contain not only harmonic components, but also signals having frequencies equal to the sum and difference of the original input frequencies. Thus, the effect of second-order distortion is the production of spurious signals. The original signals are also present in the output in undistorted form. This is shown in Fig. 6-9. The seriousness of this form of distortion depends on whether one or more of the spurious signals fall in a channel that is in use. In a small system carrying only a few tv channels, second-order distortion is often not serious. In a modern system carrying midband and super-band channels, second-order distortion is very serious because the spurious signals will fall in channels that are used. This accounts for some of the problems that are encountered when a small system is expanded to carry additional signals.

The next type of distortion that we will discuss is called third-order distortion. This means that the third term on the right side of our equation is large enough to be significant. The addition of one more term to our equation looks innocent enough, but it has very serious implications. The third term means not only that additional spurious signals will be generated in the amplifier, but also that the modulation from one or more of the original input signals will appear on the other input signals. This is called *cross modulation*. Whereas with second-order distortion, the original signals were present in the output in undistorted form along with the spurious signals, with third and higher orders of distortion, the original signals that appear in the output are carrying each other's modulation. This cannot be removed.

There are several ways in which cross modulation will degrade a television picture. Even if only one channel is carried, the picture and sound carriers will cross-modulate producing buzz in the sound and "sound lines" in the picture. In order to minimize this effect, the sound carrier is usually carried through the system at a much lower level than the picture carrier. When two or more channels are carried, the sync pulses will cross-modulate, producing the "windshield wiper" effect in which a vertical bar moves back and forth across the picture.

Usually, second- and third-order distortion products are large enough to be the limiting factor in application of an amplifier, but as better transistors are developed, these terms become smaller and higher-order terms become more significant. In general, the higher the order of the term of the equation, the more spurious signals will be developed. In a system carrying many channels, thousands of spurious signals will be developed. Usually, a computer analysis is used to identify the various signals. One such computer analysis shows that a surprising number of spurious signals fall into channel 4. This bears out the complaint of cable tv technicians that channel 4 is often the hardest channel on which to get a good picture.

DISTORTION IN CASCADED AMPLIFIERS

From the earlier discussion of how cascading several amplifiers degrades the noise performance of a system, we might suspect that cascading amplifiers would also degrade the distortion situation. This is indeed true. In fact, although the distortion products are not random in nature like thermal noise, they do tend to be uncorrelated. This means that they will add up on a power basis, just the way noise does.

To overcome the effects of noise in cascaded amplifiers, we kept the signal above a certain level. This will not work in combatting the effects of distortion because more distortion is produced at high signal levels. To reduce distortion, we must keep the signal below a certain level.

Cable tv amplifiers are rated in terms of their *output level*. This is the highest level that the signal can have at the output of an amplifier without the distortion exceeding the manufacturer's specifications. If only one amplifier were used in a system, we would operate it at this level. When more than one amplifier is used, we must derate the output level; that is, the signal level must be reduced at the outputs of the amplifiers so that the distortion products from several cascaded amplifiers will not be greater than the amount specified by the manufacturer for a single amplifier.

Since the distortion products add on the basis of their power, the amount by which we must reduce the output level is 10 log n, where n is the number of amplifiers in the system. This means that we can calculate the maximum permissible signal level in a system containing n amplifiers from the equation

$$P_s = P_a - 10 \log n$$

where,

P_s is the maximum permissible signal level in the system,
P_a is the output power-level rating of a single amplifier,
n is the number of amplifiers in the system.

The maximum permissible signal level in a system consisting of several amplifiers having a gain of 20 dB and an output level rating of 50 dBmV is plotted in Fig. 6-10. Note that in some ways this plot is the opposite of that of the minimum permissible signal level given in Fig. 6-7.

The fact that the minimum permissible signal in a system decreases as the system becomes longer, while at the same time the maximum permissible signal decreases, leads to the rather unexpected conclusion that there is an optimum value of amplifier gain that will lead to the longest possible system. This will be explored in more detail under amplifier spacing.

In Fig. 6-11, the plots of Figs. 6-7 and 6-10 are combined. It can be seen that, with the values assumed, the longest possible system

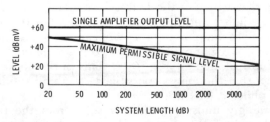

Fig. 6-10. Derating of output level in cascaded amplifiers.

is one having a total length of a little less than 2000 dB. This length is obtained with a signal level at the output of each amplifier of about 30 dBmV. Note, however, that this would not be practical because of unavoidable changes in the parameters of the system. If the signal level were to decrease slightly, the signal-to-noise level at the output of the last amplifier in the system would be below our specified minimum. In the same way, if the signal level were to increase slightly, the distortion would be excessive. This latter

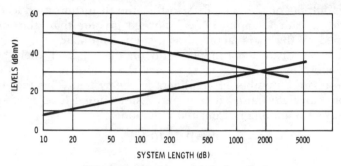

Fig. 6-11. Maximum system length.

parameter is often referred to as the *signal-to-distortion* ratio. It is the difference in decibels between the signal level and the maximum permissible signal level.

AMPLIFIER GAIN AND SPACING

From the preceding discussions of noise and distortion, we see that there are limits to both how low and how high the signal level may be in any given system. If the signal level drops below the lower limit, the picture will be degraded by noise. If the signal goes higher than the upper limit, the picture will be degraded by distortion. The difference between the upper and lower permissible signal levels is called the *dynamic range* of the system.

Since the two limits tend to coincide as the system becomes longer, we can conclude that there is an optimum gain of the amplifiers in a system that will give the longest possible system. By assuming that the noise figure and the rated output level of the amplifiers do not change with amplifier gain, we can calculate the optimum amplifier gain. We will omit the detailed mathematical analysis, but the results are plotted in Fig. 6-12. This curve shows the rather amazing fact that the longest possible system will result if the gain of each amplifier is 8.69 dB.

This seemingly unusual number results from the fact that the decibel is based on the system of common logarithms, which has ten as its base. This definition of a base of a system of logarithms is purely arbitrary and has no basis in natural growth. In any system where the change in a quantity is proportional to its value, the number $e = 2.718$ appears. A system of logarithms, called natural logarithms, based on e rather than 10, would yield an optimum gain of 1. This is the same as saying that the gain of the optimum amplifier is equal to e, which is obvious from the equation

$$20 \log e = 8.69$$

Returning to Fig. 6-12, the plot shows that there is nothing to be gained by making the gain of a cable tv amplifier as high as possible. Another amazing conclusion that can be drawn is that the optimum gain is independent of both amplifier noise figure and output level, as long as these are independent of gain.

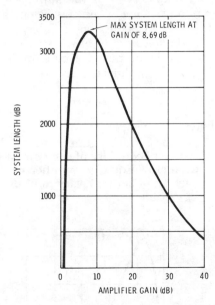

Fig. 6-12. System length vs amplifier gain.

In practice, there are several reasons why the amplifier gain is greater than the optimum value we have derived. In the first place, such a low-gain amplifier would of necessity be a single-stage affair. In a single-stage amplifier, the designer can design for minimum noise figure or maximum output power level, but not both in the same amplifier. Lowest noise figure results from using a small transistor with small currents, whereas maximum power output level is obtained by using a large transistor with a comparatively large collector current. Furthermore, the system using amplifiers with a gain of 8.69 dB made no allowance for variations in signal or system parameters. Although compensation schemes are used to minimize these variations, they will inevitably be present to some extent.

Most modern cable tv amplifiers use at least two stages. The input stage is designed to provide a low noise figure, and the output stage is designed to provide a high output level with minimum distortion. This scheme cannot be carried out completely because the input stage will always contribute some small amount of distortion and the output will always contribute some small amount of noise.

Still another consideration in determining the gain of the amplifiers is the economics of the complete system. A smaller number of amplifiers with gains higher than the optimum value will restrict the length of the system but will reduce the cost. Increasing the power output level of the amplifiers will increase the power consumption and hence the cost of operation of the system.

One way of taking unavoidable variations into consideration is to define the *uncertainty* of the system in decibels. This is the amount by which the output of each amplifier may vary, multiplied by the number of amplifiers. The optimum gain of each amplifier then is given by:

$$A_{opt} = 8.69 + U_s$$

where,

A_{opt} is the optimum gain of the amplifier in decibels,
U_s is the uncertainty of the gain of the amplifiers in the system in decibels. (Equal to number of decibels by which the gain of each amplifier may vary multiplied by the number of amplifiers in cascade.)

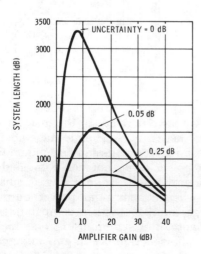

Fig. 6-13. Optimum amplifier gain with amplifier uncertainty.

A plot of possible system length versus amplifier gain for various values of uncertainty in the system is given in Fig. 6-13. The top curve is the optimum gain of 8.69 dB that we found in the idealized case. The gain in this case is rather critical. The system will be shortened considerably by small changes in gain. Even a small amount of uncertainty reduces the maximum possible length considerably, but the curve becomes much flatter, which means that variations in gain are not nearly as important.

Modern cable tv amplifiers usually have a gain of 18 to 25 dB. The noise figure is always as low as practical, and the output level is made as high as practical.

TILT AND SLOPE

It has been stated that the attenuation of a coaxial cable is not the same at all frequencies. It varies very nearly with the square root of frequency. Thus, the attenuation at channel 13 will be much greater than that at channel 2. Since the object of the system is to deliver signals to the subscriber's home that are all at about the

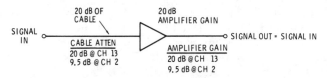

Fig. 6-14. Cable losses vs amplifier gains.

same level, the amplifiers in the system contain equalizers that will compensate for the nonuniform attenuation of the cable. Fig. 6-14 shows a 20-db length of cable followed by a 20-dB amplifier. These figures acutally apply only at the highest frequency of interest—in this case, channel 13. The loss of the cable at channel 2 is only about 9.5 dB. So that signals of all frequencies of interest will leave the amplifier at the same level at which they entered the cable, the gain of the amplifier must increase with frequency. Its gain-versus-frequency curve must match the attenuation curve of the cable. The system is then said to be *flat*. This variation in amplifier gain with frequency is referred to as *slope*, and it is a property of the amplifier.

Fig. 6-15 shows the level versus frequency at both the input and output of the amplifier for different types of input signals. In Fig. 6-15A, the input to the cable is flat at all frequencies of interest, and so is the output of the amplifier, but at the input of the amplifier, the signal level decreases with frequency. This variation in amplitude with frequency in the *system* is called *tilt*. In Fig. 6-15A the input

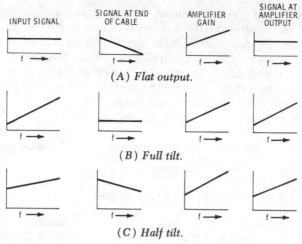

INPUT SIGNAL SIGNAL AT END OF CABLE AMPLIFIER GAIN SIGNAL AT AMPLIFIER OUTPUT

(A) *Flat output.*

(B) *Full tilt.*

(C) *Half tilt.*

Fig. 6-15. Possible modes of system operation.

to the amplifier is fully tilted. A system operating in this mode is said to have a *flat output.*

The opposite situation is shown in Fig. 6-15B. Here the input to the system and the signal at the output of the amplifier are fully tilted. The signal at the input to the amplifier is flat. This mode of operation is referred to as *full tilt.* Fig. 6-15C shows a situation somewhere between the other two that is called *half tilt.*

In the past, the terms slope and tilt have been used loosely and interchangeably. The control in an amplifier that changes its gain with frequency has been called both a slope control and a tilt control. The latest NCTA standards clearly specify that slope is a property of an amplifier describing its variation in gain with frequency. Tilt is a property of the system, and the amount of tilt is specified at the outputs of the amplifiers. All cable tv amplifiers have circuits and adjustments to control the slope.

PRACTICAL CABLE TV AMPLIFIERS

Amplifiers used along the cable of a system are usually classified into three categories. *Trunk,* or *mainline,* amplifiers are spaced along the main trunk lines of the system to compensate for cable attenuation. *Bridger* amplifiers are located either in the same case with the trunk amplifiers or separately and are used to take a signal off the trunk line for distribution in feeder cables to the subscriber's neighborhood. *Line extenders* are amplifiers that are used along feeder systems where required. Some amplifiers can be used for any of the three purposes.

Trunk amplifiers have the most stringent specifications simply because many of them are cascaded in a long system. Line extenders can get by with lesser specifications where only one or two are used on a feeder line. Since amplifiers are located throughout the system, they are mounted in weathertight cases, as shown in Fig. 6-16, which allows them to be mounted on poles, pedestals, or even in manholes in underground systems. Connectors are provided to permit connection to the cable system with a minimum of mismatch.

Fig. 6-16. Typical amplifier housing.

Earlier cable tv amplifiers used vacuum tubes, and many of these amplifiers are still in service. Modern amplifiers all use solid-state circuits—either transistors, integrated circuits, or both. Early transistor amplifiers were inferior to vacuum-tube amplifiers in that their dynamic range was usually not as great. They became popular principally because their power requirements were so much lower. The transistor amplifier made it possible to power amplifiers by sending the power along the same coaxial cable that carried signals. The power requirements of vacuum-tube amplifiers were so high that it was necessary to make connection to a power line at every amplifier location. Because of their low power requirements, transistor amplifiers made shorter cable spacings economically practical.

Newer solid-state amplifiers surpass vacuum-tube amplifiers in almost every respect.

Fig. 6-17 shows a block diagram of a cable tv amplifier. Two separate stages of amplification are used. In the design of this type of amplifier, an attempt is made to have the first stage determine the noise contribution and the output stage control the overload level. This is not entirely possible because the gain of both stages is low compared with the gains of other types of amplifiers. The output stage will contribute some noise, and the input stage will contribute some distortion.

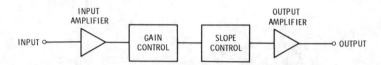

Fig. 6-17. Block diagram of a cable tv amplifier.

At the input to the amplifier, an attenuator and equalizer are used to establish the proper input signal levels. Often, temperature compensation is also introduced here. Between the stages additional equalization is included, and usually a control is provided to establish the slope of the amplifier. A small range of gain control is generally provided to compensate for irregularities which, as we will see later, usually show up as being equivalent to an error in amplifier spacing.

In systems that carry signals only on regular television frequencies between channels 2 and 13, second-order distortion is usually not a serious problem, if it is within reasonable limits. The spurious sum and difference frequencies fall in portions of the spectrum that are not used to carry picture signals. On systems that carry additional signals on the midband channels between channels 6 and 7 and on the channels above channel 13, this is not true. The spurious signals will then fall into channels that are in use. To minimize second-order distortion, push-pull amplification is often used in all but very low-level stages.

Fig. 6-18 shows a simplified block diagram of a push-pull amplifier. At the input to the amplifier, the signal is split into two separate components that differ in phase by 180 degrees. The two out-of-phase signals are amplified separately and are combined in an output transformer in such a way that they reinforce each other. Second-order products generated in the amplification process appear in phase at the output transformer, and they are combined in such a way that they cancel each other. The amount of cancellation that takes place depends on how well balanced the push-pull stages are. Cancellation is not really complete because the amplifier and transformers cannot be perfectly balanced. It is, however, sufficiently

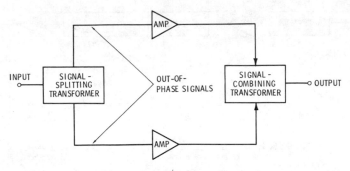

Fig. 6-18. Push-pull amplifier.

effective that second-order distortion is usually not a limiting factor in the design of a midband system.

TEMPERATURE COMPENSATION

In a practical system, it is necessary to use some method of compensation for changes in the attenuation of the cable that are caused by such things as temperature variation. The attenuation of most coaxial cables used in cable tv will increase approximately 0.1% for each degree Fahrenheit of temperature rise. This figure superficially may seem small, but in a long system, this much variation of attenuation is absolutely unacceptable. Suppose, for example, that a system were 1000 dB long. The total cable loss would be 1000 dB, and the total amplifier gain would be 1000 dB. In temperate latitudes, the ambient temperature can be expected to vary over 100 degrees throughout the year. This means that the attenuation of the cable may change by $1000 \times 0.001 \times 100 = 100$ dB. Since the amplifier gain will remain constant over this temperature range, the signal at the subscriber's home would vary 100 dB from the coldest day in winter to the hottest day in summer.

To compensate for this and other small changes in the system, it is common practice to use some form of automatic gain control (agc) or automatic level control (alc). In modern systems, automatic slope control is also provided.

The simplest form of alc consists of transmitting a pilot carrier along the system with the television signals. About every third amplifier along the system includes a circuit that filters out the pilot carrier, detects its level, and controls a variable attenuator that adjusts the gain of the amplifier.

A more elaborate system is shown in Fig. 6-19. Here two pilot carriers—one at 73 MHz and one at 271 MHz—are carried throughout the system. In the amplifier, circuits are provided to detect

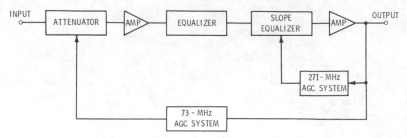

Fig. 6-19. Block diagram of an amplifier with agc and asc.

separately the level of each pilot carrier and to adjust the gain of the amplifier at the two separate frequencies. This system adjusts not only the gain of the amplifier but also its slope. Amplifiers of this type are usually spaced about every third amplifier along a trunk.

SIGNAL SPLITTING

It is rarely possible to design a system in which all of the subscribers are located close to a straight trunk cable, and for that matter, it is not even desirable. It is necessary to branch the signal from the headend along different paths at many points in the system. This splitting of the signal must be done in a way that will introduce a minimum amount of mismatch at the junction. The splitting may be accomplished by passive components that are similar to a directional coupler. Often this splitting arrangement is included within the same case as the amplifier.

If the signal leaving the amplifier is, to be split into two paths with equal signal levels, a simple 2-way splitter can be used at the output of the amplifier. If unequal division is required, a directional coupler can be used. There is a definite limit to the amount of splitting that can be done with passive components without upsetting the spacing of the system. The reason is simply that there is only so much signal leaving the amplifier. What is diverted to a

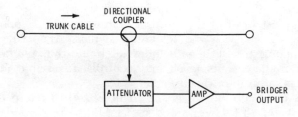

Fig. 6-20. Bridging amplifier.

different path is taken out of the main trunk. A more suitable way to get signals out of the trunk is to couple a small amount of signal out of the cable through a directional coupler and feed it directly to an amplifier that will bring the signal to the desired level. An amplifier for this purpose is called a bridging amplifier, or simply a bridger.

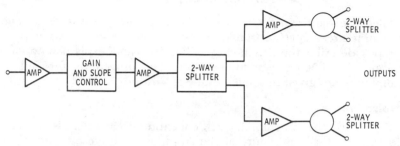

Fig. 6-21. Amplifier designed for four equal outputs.

Fig. 6-20 shows a block diagram of a bridging amplifier. It may be housed in the same case with a trunk amplifier or be packaged separately. Many elaborate arrangements are possible for particular requirements. Fig. 6-21 shows an arrangement that will provide four output signals, each having the same level as the input signal. The arrangement uses amplifier stages to bring the level to the desired value and to provide further isolation between the outputs.

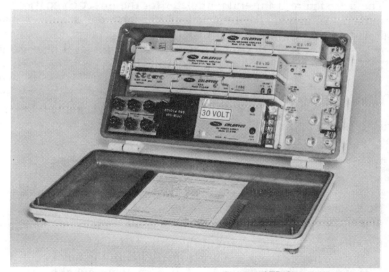

Courtesy AEL Communications Corp.

Fig. 6-22. A cable tv amplifier utilizing modular construction.

107

Modular packaging, similar to that shown in Fig. 6-22, is very popular in modern amplifiers. This approach makes it possible to include many different amplifiers, splitters, and attenuators in the same case to meet a particular system design objective.

AMPLIFIER SPECIFICATIONS

Because requirements for cable tv amplifiers are rather unique, manufacturer's specifications are also unique. The following paragraphs describe the common specifications and explain how each parameter influences system operation. Test procedures for cable tv amplifiers are given in a later chapter.

Noise Figure

The noise that an amplifier will contribute to a system is specified in terms of the noise figure of the amplifier. The noise figure, F_a, is in decibels and is given by the expression:

$$F_a = 10 \log \frac{S_i/N_i}{S_o/N_o}$$

where,

S_i is the input signal level,
S_o is the output signal level,
N_i is the input noise level,
N_o is the output noise level.

The noise figure is a useful parameter for the direct comparison of amplifiers. An amplifier with a noise figure of 8 dB is quieter than one with a noise figure of 12 dB. To apply the concept to a system, it is useful to convert the noise figure to a definite level of noise in dBmV. If the input of an amplifier is terminated in a resistance equal to its input impedance, it will act as a noise source. The noise level at the output will consist of the thermal noise at the input degraded by the noise contribution of the amplifier and increased by the gain of the amplifier. Thus, the noise level, N_o, at the output of the amplifier is:

$$N_o = -59.1 + F_a + A$$

where,

-59.1 dBmV is the thermal noise in a 4-MHz television channel,
F_a is the noise figure of the amplifier in dB,
A is the amplifier gain in dB.

What this equation says is that at the output of the amplifier we will have the inevitable thermal noise, together with the noise contribution of the amplifier, both amplified by an amount equal to the gain of the amplifier.

Frequency Range

The specified frequency range is merely the range of frequencies across which the amplifier is designed to operate. It usually starts near 50 MHz and may go as high as 300 MHz.

Flatness

It is important that the gain of an amplifier connected to the desired length of cable be the same at all frequencies. The flatness of an amplifier is specified by the decibel variation over the frequency range of interest. It only applies when the associated length of cable is equal to the gain of the amplifier.

Gain

Amplifier gain is specified in decibels at the frequency of the highest-frequency channel. It is numerically equal to the recommended amplifier spacing. Most amplifiers have a provision for varying the gain over a range of a few decibels to compensate for inevitable errors in actual spacing, or other errors that are equivalent to errors in spacing. Usually, changing the amplifier gain will also change its slope, so a control is provided to adjust the slope. Some amplifiers have slope-compensated gain controls.

Output Level

The output level of an amplifier, specified in dBmV, is the highest level that can be obtained without distortion exceeding the specified limits. It is specified for the frequency of the visual carrier of the highest-frequency channel used.

Output level is specified in dBmV of peak envelope power, not average power. Peak envelope power is the power developed by an unmodulated carrier having a peak amplitude equal to that of the modulated carrier. Since there is an impedance of 75 ohms at any point in a cable tv system where output level would be measured, the output level in dBmV is given by:

Output Level = 20 log (rms output in millivolts at peak modulation)

The output level of some older amplifiers was specified in terms of output capability, and at one time there was a standard for this quantity. The concept was vague, however, because it made no distinction between various types of distortion and was very difficult to measure repeatably. It no longer appears in the NCTA standards.

Distortion

From the point of view of providing high-grade signals, the distortion specification of an amplifier is as important as the noise specification. The current NCTA standard for amplifier-distortion characteristics recognizes two types of distortion—cross modulation and spurious signals. Hum and noise are not included in the distortion specification.

It was pointed out earlier that second-order distortion produces spurious signals which add to existing signals but do not corrupt the existing signals. Third and higher orders of distortion also produce spurious signals, but, in addition, they actually impose the modulation of one carrier on another. This phenomenon is known as cross modulation.

It is important that all operating conditions of an amplifier be specified when talking about distortion. An amplifier that would carry four or five channels without difficulty might be totally unuseable in a system with many more channels.

Hum Modulation

Hum modulation is the amount in decibels by which any hum modulation of the carrier is below the carrier level.

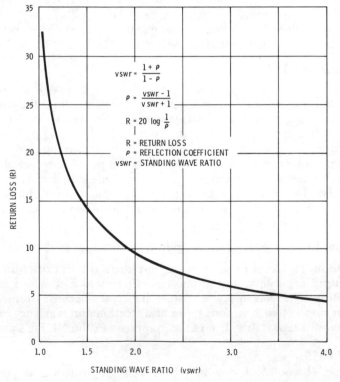

Fig. 6-23. Relationship between return loss and vswr.

Input and Output Mismatch

Usually the actual input and output impedances of a cable tv amplifier are not even mentioned in the specifications because almost without exception they are designed to work with 75-ohm coaxial lines. What is specified is the amount of mismatch. This is specified in decibels of return loss. Return loss was described in an earlier chapter. The reason for using this parameter as a measurement of mismatch rather than some other quantity, such as standing-wave ratio, is that it lends itself to comparative measurement in broadband systems. Fig. 6-23 shows the relationship between return loss and standing-wave ratio.

Power Requirements

Operating power for a modern solid-state cable amplifier is obtained from an ac voltage which is carried through the cable along with the television signals. Both 30-volt and 60-volt systems are in common use. Earlier vacuum-tube amplifiers with their higher power requirements were usually operated from a 120-V ac line.

The power specification for an amplifier gives the operating voltage and current. Usually, some description is given of the arrangement that will permit operating power to be coupled through the amplifier to the output side of the cable. More will be said about powering cable tv amplifiers in a later chapter.

Chart 6-1. Typical Cable TV Amplifier Specifications

Bandwidth	50–270 MHz
Flatness	±0.25 dB
Output Level	32 dBmV
Full Gain	23 dB
Gain Control Range	0–7 dB
Slope Control	0–8 dB
AGC Operation	
73-MHz Output	20 dBmV
271-MHz Output	24 dBmV
AGC Regulation (8-dB Cable Change)	±0.5 dB
Slope Regulation (8-dB Cable Change)	±0.5 dB
Input Match	18 dB Return Loss
Output Match	16 dB Return Loss
Noise Figure	
At 270 MHz	10 dB
At Channel 2	12 dB
Cross Modulation	−90 dB
Intermodulation (Spurious)	−80 dB
Hum Modulation	−70 dB
Power Requirements	
Current at 30 V rms	1.2 A
Current at 60 V rms	0.6 A

Automatic Operation

If an amplifier has automatic level control or automatic slope control, the frequency of the pilot carrier or carriers is given. The range of automatic control in decibels is given for changes in either input or output signal level.

Other Specifications

If an amplifier is designed for use in two-way systems, the frequency range and characteristics of the upstream channels are given. Other specifications might include the types of connectors, mechanical dimensions, and weight.

Chart 6-1 shows a typical set of cable tv amplifier specifications.

CHAPTER 7

The Headend—
Antennas and
Propagation

The function of the headend of a cable tv system is twofold—to pick up signals off the air and to process them for transmission along the cable distribution network. In this chapter we will be concerned with the first of these two functions. The selection of a suitable site for the headend and the choice of antenna type depend on the signal environment, that is, on the number of signals available and their strengths. This, in turn, depends on the distance to the broadcast stations and the mode of propagation by which the signals reach the location.

It is well to bear in mind that the function of the antenna is not only to provide good reception of desired signals but, equally important, to discriminate as much as possible against undesired signals on the same or adjacent channels. This is often the more stringent requirement.

WAVE PROPAGATION

In understanding the way in which radio waves are propagated through space, it should be realized that the velocity of propagation of a radio wave is a function of the dielectric constant of the medium through which it passes, with free space having a dielectric constant of 1. When a radio wave passes through empty space, it will travel in a straight line. When passing through air, which has a dielectric constant of slightly greater than 1, it will still travel in a straight line but at a slightly lower velocity. The atmosphere surrounding the earth is not homogeneous; it has slight variations which will be described later. When a radio wave passes through a nonhomogeneous medium, it will no longer travel in a straight line, but will be reflected or refracted (bent), or both. The fact that the atmosphere and ionosphere are not homogeneous is what makes long-distance radio communication possible.

Fig. 7-1 shows a sketch of the environment surrounding the earth. Only those features that affect the propagation of signals at frequencies used for television are labeled. The layer extending from the surface of the earth to an altitude of about six miles is called the *troposphere*. It is in this layer that almost all of our weather, such as rain, snow, wind, and storms, occurs. Most of the moisture in the atmosphere is in this layer. Because of the fact that the amount of moisture in the troposphere varies from one place to another and with time, the dielectric constant varies considerably. When a radio wave passes through these variations of dielectric constant, it is refracted or bent.

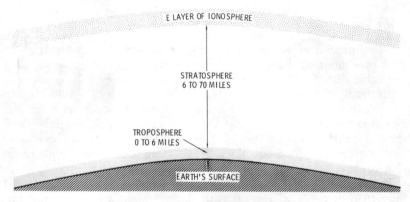

Fig. 7-1. Layers of the atmosphere.

Above the troposphere is a layer where the temperature is nearly constant and there is very little moisture. This layer, extending from the troposphere to an altitude of about 70 to 100 miles, is called the *stratosphere*. Here, the dielectric constant does not vary, and this layer has very little effect on propagation of television signals. Occasionally, when there is a visual display of the aurora (northern lights), due to magnetic storms, ionization in this layer will affect television signals, but this is rare and not predictable.

Above the stratosphere, at an altitude between 70 and 100 miles above the earth, is the so-called *E-Layer* of the *ionosphere*. Most of the time this layer is not ionized strongly enough to affect television signals. Occasionally, however, there are patches of extremely high ionization called *sporadic E*. When sporadic E occurs, it will reflect signals up to 100 MHz or sometimes higher. Because of the height of the phenomenon, the closest point at which the reflected signal will reach the earth will be about 500 miles from the transmitter location. This mode of propagation is of no interest in the propagation of desired signals, but it often explains the presence of interfering signals on the low-band channels.

In general, because of the layers described above, television signals may reach the headend of a cable tv system through any of four modes of propagation—line of sight, beyond the horizon, tropospheric scatter, or sporadic E.

Line-of-Sight Propagation

Line-of-sight propagation is exactly what the name implies—propagation of a signal from one antenna to another above the horizon. If there were no obstructions, the transmitting antenna would be visible from the headend antenna. In this mode of propagation, the signal usually reaches the receiving antenna through two paths, as shown in Fig. 7-2. One path is direct from one antenna to the other. The other is a reflected path in which the signal bounces off the ground somewhere between the two antennas.

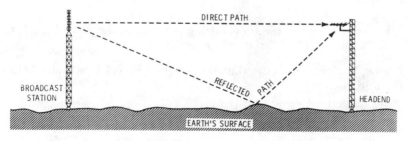

Fig. 7-2. Direct and reflected signal paths.

The signal strength at the receiving antenna is the vector sum of the signals arriving via the direct and reflected paths. Since the reflected signal travels over a longer path than the direct signal, the two are usually not in phase. They may either reinforce each other or tend to cancel, depending on their phase relationship.

The lengths of the two propagation paths will vary as the receiving antenna is raised or lowered. This will change the relative phases of the two signals so that the signal strength at the receiving antenna will vary periodically as its height is changed. This is shown in Fig. 7-3.

115

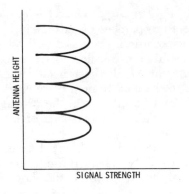

Fig. 7-3. Variation of received signal strength as antenna height changes.

It is interesting to note that the reflected wave will disappear if the receiving antenna is located exactly on the horizon. At this point the reflected signal can be ignored.

Beyond-the-Horizon Propagation

Because of the curvature of the earth, there is a limit on how far two antennas of given heights can be separated and still be above the horizon. This distance increases as the antennas are raised above the earth and is given by the equation

$$d = \sqrt{2h_T + 2h_R}$$

where,

d is the distance between the two antennas in miles,
h_T is the height of the transmitting antenna in feet,
h_R is the height of the receiving antenna in feet.

A receiving antenna located at point A in Fig. 7-4 is actually below the horizon.

Fig. 7-4. An example of the radio horizon.

It was pointed out earlier that radio waves will not travel in straight lines in a nonhomogeneous medium but will be bent. The troposphere is not uniform, even in the absence of severe weather conditions. There is normally a decrease in temperature, pressure, and water-vapor content at higher altitudes. This tends to bend the radio waves slightly toward the surface of the earth. As a result, the signals propagate about 30% farther than would be expected from line-of-sight considerations. For this reason, we say that the *radio horizon* is about 30% farther away than the visual horizon.

Over-the-horizon propagation can be conveniently calculated by considering the radius of the earth to be about 30% greater than it really is. This method is sometimes referred to as using a 4/3 earth radius.

When atmospheric conditions are such that there are abrupt changes in temperature and water-vapor content from one point to another, there is additional bending, or refraction. This effect is particularly noticeable in summer and is more prevalent in certain parts of the country, such as the west coast regions of the United States.

Tropospheric Scatter Propagation

Another way in which high-frequency signals are propagated is called *tropospheric scattering.* Scattering is a familiar phenomenon with light. Everyone has seen the glow of a city's lights or the beam of a searchlight in the sky even though the city or the searchlight is actually below the horizon. The reason that the glow in the sky is visible is that some of the light is scattered back toward the earth.

Television signals are scattered in the same way when they encounter patches in the troposphere where the temperature or water-vapor content differs from that in the rest of the region. These little patches are always present in all kinds of weather but are more plentiful at some times than at others.

Signals that are propagated by tropospheric scattering are not constant but are subject to types of fading. In one type, there are changes that take place over a period of hours or even days. These are due to gradual changes in temperature or water-vapor content of the troposphere. In a second type of fading, tropospheric-scatter signals are subject to a rapid fading in which the signal varies from maximum to minimum strength in only a few seconds. This is called *multipath fading* and is due to the fact that signals scattered by different patches of abnormal temperature or moisture alternately reinforce and cancel each other.

Sporadic E Propagation

As mentioned earlier, sporadic E often accounts for the presence of interfering signals that are normally absent. Because of the height of the E-layer, the interfering station will always be at least 500 miles away.

Auroral Propagation

A severe magnetic storm that produces a visual auroral display will sometimes produce interfering signals. Usually all of the modulation on a signal will be destroyed and the interfering signal will be recognizable only as higher than normal noise. In the United

States, auroral signals have the strange property that they tend to arrive from the north, regardless of the actual direction of the interfering station.

Ghosts

The familiar form of interference called a *ghost* occurs when a signal actually interferes with itself. Ghosts may be introduced at many different points in a system, but here we are interested in ghosts that are actually picked up by the antenna. Fig. 7-5 shows the geometrical situation that leads to ghosts. The signal from the television broadcast station reaches the antenna at the headend by two different paths. One path is direct, and the other involves reflection of the signal from some large object in the area such as a water tower. In any well-selected and properly installed headend, the ghost will always be weaker than the direct signal. As a result, the direct signal will take command of the synchronizing circuits in the receiver, and the ghost signal will be displayed on the screen displaced to the right of the regular picture. This ghost to the right of the picture is known as a *trailing ghost*. The fact that it is trailing means that the weaker of two signals arrives at the screen of the receiver later in time than the other signal.

The amount by which a ghost is displaced on a screen is related to how much longer the reflected path is than the direct path and is given by the equation:

$$d = \frac{rw}{cT_H} = \frac{rw}{52,200}$$

where,

d is the displacement in inches of the ghost on the picture-tube screen,

r is the difference in feet between the lengths of the reflected and direct paths,

c is the velocity of propagation, about 985 feet per microsecond,

T_H is the duration of the horizontal scanning line (53 μs),

w is the width in inches of the picture.

This expression can be used with a detailed map of the area to identify which structure is actually causing a ghost. The formula can also be used to determine the length of a reflected path that will cause enough delay to make a ghost objectionable.

If the resolution of the picture is 450 lines, the width of a barely discernible picture element on the screen will be 450 divided into the width of the screen. On a 10-inch screen, this means that for a ghost to be noticeable the reflected path must be at least 115 feet longer than the direct path.

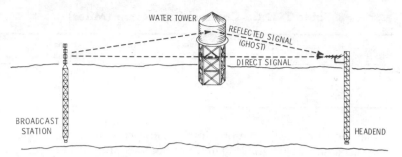

Fig. 7-5. Reflections cause ghosts.

Ghosts can usually be avoided at the time of system installation by proper selection and orientation of antennas. Nevertheless, at any time after system installation, the construction of new buildings, bridges, water towers, or similar structures may cause new ghosts to appear.

ALLOCATIONS

The allocation of channels to television broadcast stations is intended to minimize both co-channel and adjacent-channel interference. For purposes of allocation, the United States is divided into

Fig. 7-6. Zones for tv station allocation.

the three zones shown in Fig. 7-6. The minimum allowable separation between stations operating on the same channel in each of these zones is shown in Table 7-1. In the event that two stations are in different zones, the minimum separation will be the smaller of the two.

Table 7-1. Co-Channel Separation (Miles)

Zone	Channels 2–13	Channels 14–83
I	170	155
II	190	175
III	220	205

Television broadcast stations in the same community are not assigned to adjacent channels. The minimum separation between two stations operating on adjacent channels is:

Channels 2 through 13	60 miles
Channels 14 through 83	55 miles

Not all television channels are contiguous in the spectrum. There are gaps between channels 4 and 5, 6 and 7, and 13 and 14, so these pairs of channels are not considered to be adjacent channels.

In most locations this system of allocation has proven effective in preventing interference in home receivers with modest antenna systerms. Cable tv systems, however, use tall towers, much higher than the average home antenna, with the result that they pick up weaker signals. In fact, minimizing pickup from undesired stations on the same and adjacent channels is often a major consideration in selecting a site and type of antenna.

Co-Channel Interference

Co-channel interference is interference from a television broadcast station operating on the same channel as the desired station. The visual effect of an interfering signal is the presence of horizontal or diagonal lines across the picture. The lines may be stationary or may move up and down on the screen. The number of lines that will appear is equal to the frequency difference between the two visual carriers divided by the frame rate. The FCC has taken advantage of this to minimize the effect of co-channel interference by requiring one of two stations that might cause co-channel interference to offset its visual carrier by 10 kHz.

The amount of interference that can be tolerated depends entirely on the subjective reaction of the viewer. The Television Allocations Study Organization (TASO) investigated the amount of interference that a large group of viewers found to be objectionable and rated the effect of interference as shown in Table 7-2. This subject is still being considered by the Cable Television Advisory Committee (C-TAC), but at present a signal-to-interference ratio of 48 dB is considered to be a reasonable objective for cable tv systems.

Adjacent-Channel Interference

Fig. 7-7 shows the location in the spectrum of the visual and aural carriers of a television channel together with the sound carrier of the lower adjacent channel and the visual carrier of the upper adjacent channel. Interference to a picture can result from a beat between any of the four carriers or between any of the carriers and

Table 7-2. Subjective Effect of Interference

Signal-to-Interference Ratio (dB)	Picture Quality
37.2	Passable
42.6	Fine
48	Excellent

any of the modulation. A very common form of adjacent-channel interference results from heterodyning of the visual carrier with the sound carrier in the lower adjacent channel. The result is a coarse interference pattern.

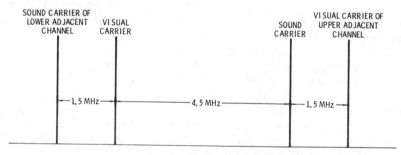

Fig. 7-7. Relationship of visual and aural carriers of adjacent channels.

Another form of adjacent-channel interference results when the picture sideband in the upper adjacent channel beats with the visual carrier in the desired channel. The predominant feature of this type of interference is the presence of a vertical bar (actually the blanking pulses) in the picture. Because the synchronizing pulses in the two pictures are not synchronized with each other, the bar will move back and forth across the picture. This is usually called the "windshield wiper effect."

The cause of adjacent-channel interference is not always easy to pin down. It may result from an interfering signal being picked up by the antenna, but it might also be introduced at other places in the system.

121

LOCATION OF THE HEADEND

It is obvious that the signal from the headend antenna should have the most favorable signal-to-noise and signal-to-interference ratios possible. For this reason, the location of the headend must be chosen with great care.

Site Selection

In selecting a site for the headend of a cable tv system, the complete electromagnetic environment at television frequencies must be known as accurately as possible. Several techniques are commonly used for site selection and evaluation. These range from elaborate computer programs, through aerial surveys, to ground-based measurements. In any case, the probable signal strength and azimuth bearing and distance to each station must be known. This includes not only stations whose signals are to be carried by the system but also stations that might cause co-channel or adjacent-channel interference.

The survey often is made on a map of the area that will show distances up to 150 to 200 miles from the site. Maps that are suitable for this purpose can be obtained from the U.S. Government Printing Office. Ordinary highway maps are often quite inaccurate and should not be used.

Calculation of expected signal strength is not a simple matter. In order to make the calculation, the transmitter power, antenna height, and antenna type must be known as well as the mode of propagation by which the signals reach the site. This is not always clear cut. Sometimes a signal will reach a site by two overlapping modes of propagation. Usually, interfering signals can be evaluated on the basis of the probability that they will not cause interference more than a small percentage of the time.

Computer site surveys are made by consultants who maintain large banks of data about all of the signals in the area.

Site Surveys

One of the best ways to determine the electromagnetic environment at a proposed site is to make a signal survey on the spot. This is usually done with portable antennas that can be raised on a temporary mast to a height of 70 to 100 feet above the ground. Measurements should be made on all vhf channels and on all uhf channels of interest, as well as on adjacent channels. Measurements should be made over as long a period of time as is practical in order to pick up as many abnormal conditions as possible. To be of maximum value, measurements should be made during all four seasons, but this is usually out of the question because of cost.

Sometimes the selection of a site is simplified by the fact that only three or four locations in an area are even available for a cable tv headend site.

In the site survey, the evaluation of the electromagnetic environment should not be restricted to television signals. It should include evaluation, to the extent practical, of all types of electrical noise in the television portion of the spectrum as well as their cause and the possibility of suppressing them. Strong signals at any frequency should be noted because they may require special consideration. A radar transmitter, even though its frequency is much higher than the television frequencies, might overload a preamplifier and cause interference unless special filtering is installed to reject the radar signals.

Miscellaneous interference caused by electrical equipment may sometimes be suppressed at the source by cooperating with the owner of the offending equipment. At the present time the electromagnetic spectrum is becoming polluted by spurious signals from transmitters and incidental radiation from noncommunications devices. Undoubtedly, if a catastrophe in communications is to be avoided, the day will come when pollution of the spectrum is considered as serious as pollution of the atmosphere. Unfortunately, that day hasn't arrived, and until then the system operator must do his best to select a site as free as possible from interference. He must also keep a watchful eye on industrial developments in the area that might later lead to interference.

Some types of interference require a great deal of imagination to anticipate. For example, a site that is normally ideal might become nearly useless on winter weekends because of ignition interference from snowmobiles.

ANTENNAS

The purpose of the antenna in a cable tv headend is to provide the best possible reception of desired signals, with maximum discrimination against undesired signals. The discrimination against undesired signals is often the more critical consideration. The ideal antenna would be one that had a high gain in one direction and practically no pickup in any other direction. A graphical representation of the directional properties of an antenna is called the *pattern* of the antenna. The pattern of our ideal antenna is shown in Fig. 7-8. Unfortunately, the narrower the beam of an antenna, the larger its physical dimensions in wavelengths. At the high uhf channels, very narrow beams can be obtained with antennas of practical size, but at the lower vhf channels, an antenna having the pencil beam shown in Fig. 7-8 would be the size of a battleship.

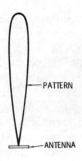

PATTERN

ANTENNA

Fig. 7-8. An ideal cable tv antenna pattern.

Gain and Directivity

The terminology used in describing antennas is based on the reciprocity theorem. This theorem states that, in general, a directional antenna will have the same pattern whether it is used to transmit or receive. That is, it will receive signals best from the direction in which it would radiate the most energy if it were used as a transmitting antenna. Thus, we run across terms like the *radiation pattern* of an antenna even though the antenna is used strictly for receiving signals and does not radiate at all. We will define the parameters of an antenna in terms of its being used to receive signals, but in some manufacturer's literature the parameters may be defined in terms of radiation. Because of the reciprocity theorem, it really makes no difference which approach is taken.

The *directivity* of an antenna may be defined as the ratio of its maximum pickup to its average pickup. The *gain* of an antenna is defined as the ratio of its maximum pickup in a given direction to the pickup of some reference antenna. Gain tells how much better an antenna will pick up a signal from a given direction than the reference antenna. If there were no losses in the antenna, the gain would be equal to the directivity. All practical antennas have some loss, so the gain will be less than the directivity.

The above definition of antenna gain involved some sort of reference antenna. A common reference antenna used for specifying antenna gain is the *isotropic* antenna. This is a hypothetical antenna that has the property of radiating or receiving signals equally well from all directions. This reference has the advantage of being easy to apply. It has the disadvantage of existing in theory only; it is impossible to build an isotropic antenna. Another common reference antenna is the simple half-wave dipole, shown in Fig. 7-9A. This antenna has some directivity as shown in Fig. 7-9B, but the pattern is very broad. The gain of an antenna referred to a half-wave dipole is the ratio of the signal picked up by the antenna to that which would be picked up by a half-wave dipole oriented to receive maximum signal.

124

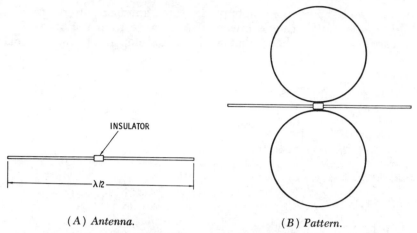

(A) Antenna.   (B) Pattern.

Fig. 7-9. Half-wave dipole antenna.

It does not matter which reference antenna is used in stating
gain, but the reference should be specified; otherwise the numbers
will be confusing. If the gain of an antenna is stated with an
isotropic antenna used as a reference, the gain figure will be about
2.15 dB higher than if it were stated with a half-wave dipole used
as a reference.

Antenna Patterns

The pattern of an antenna is usually measured on a test range
(Fig. 7-10). Here the antenna under test is rotated through 360
degrees while receiving a signal from a transmitting antenna

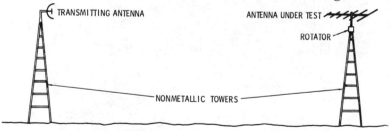

Fig. 7-10. A typical antenna test range.

located a fixed distance away. The output of the antenna under test
is plotted as a function of angle. There are several ways in which the
results of this measurement can be plotted. The coordinates may
be either polar or rectangular, and the output may be measured
linearly (in terms of voltage), in terms of power, or in logarithmic
units.

125

Fig. 7-11 shows three plots of the same antenna pattern. In Fig. 7-11A, the radial scale is linear. The distance from the origin is a measure of the voltage that will be developed at the antenna

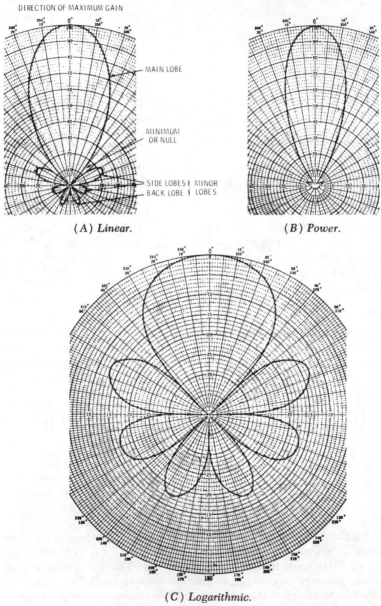

(A) Linear.

(B) Power.

(C) Logarithmic.

Fig. 7-11. Plots of an antenna pattern.

terminals at various angles. In Fig. 7-11B, the radial scale is proportional to the power developed across the antenna terminals. Finally, in Fig. 7-11C the radial scale is proportional to the change in output of the antenna in decibels.

Note that in the decibel plot the details of the pattern are much easier to discern. Furthermore, the gain in decibels can be used directly in calculations.

Several important parts of the pattern are labeled in Fig. 7-11A. The large protrusion is called the *main lobe*. It is oriented in the direction from which maximum pickup is desired. The smaller protrusions are called *minor lobes*. Those oriented toward the sides of the pattern are also called *side lobes*, and those oriented toward the rear are called *back lobes*. These minor lobes exist because of limitations in the antenna. They are more pronounced in vhf antennas where the size of the antenna is smaller in terms of wavelengths because of practical considerations.

The deep indentations in the pattern are called *nulls* or *minima* in the pattern. Strictly speaking, a null refers to an area where there is no pickup, but any of the minima are usually called nulls.

In the early days of cable tv, gain was the important consideration in selecting an antenna. There were few co-channel stations on the air at the time. Now, gain is not as important as overall directivity of the pattern. If a co-channel station is in a diametrically opposite direction from a desired station, it is usually better to use a low-gain antenna with no back lobe than a higher-gain antenna with a substantial back lobe.

It is common practice to select an antenna pattern that has a null in the direction of co-channel and/or strong adjacent-channel stations. For good co-channel suppression, the null should be as deep as possible. A 20-dB null will give marginal protection, but a 40-dB null would be ideal. However, this is rarely obtainable at the lower vhf channels where co-channel interference is most serious.

The antenna pattern does not need to be plotted on polar coordinates. Fig. 7-12 shows a pattern plotted in both polar and rectangular coordinates. In Fig. 7-12B, the horizontal scale is in angular degrees, and the vertical scale may be linear, power, or logarithmic, as in the polar plot. The rectangular scale is harder to become familiar with because it does not have the spatial coincidence that the polar plot has. However, with the rectangular plot the angular scale can be expanded as much as desired. This makes it possible to make plots that show much more detail.

The actual physical pattern of an antenna is three dimensional, as shown in Fig. 7-13. The plots we have been considering are the pattern in a horizontal plane. It is also possible to plot the pattern in a vertical plane. The plots are often identified in terms of the polarization of the antenna. Television signals are horizontally polarized; that is, the electric field, or E vector of the field, is horizontal. For this reason, the horizontal plot is often called the E-plane

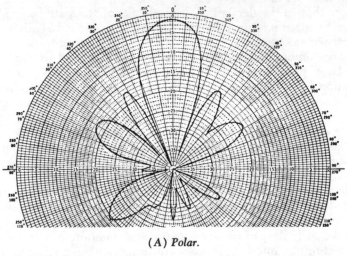

(A) Polar.

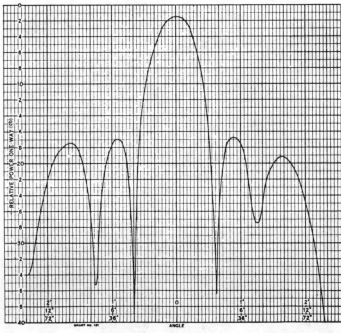

(B) Rectangular.

Fig. 7-12. Two types of antenna plot.

plot. Likewise, with horizontal polarization the magnetic field, or H vector, is oriented vertically. Thus, a vertical pattern is called an H-plane pattern.

Frequently, the vertical pattern of an antenna is ignored. It is important, however, because ghosts usually arrive at the antenna from a vertical angle. Another advantage of a narrow pattern in the vertical plane is that it will tend to discriminate against noise from distant thunderstorms.

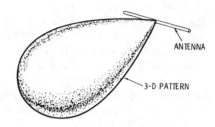

Fig. 7-13. A three-dimensional view of an antenna pattern.

ANTENNA

3-D PATTERN

It is well to bear in mind that antenna patterns are measured on a range where there is nothing to disturb the pattern. The antenna is mounted on a wooden or plastic support that will have little effect on the pattern. In operation, if the antenna is mounted on a metal tower or secured to a wooden tower with metal supports, the pattern may be changed considerably. This explains why many antennas have nearly ideal patterns when tested on a range but give a mediocre performance on a metal tower. The amount by which the pattern will be changed depends on the magnitude of the lobes oriented toward the rear where the tower is located. If the back lobes are low, the antenna is not very sensitive to what is mounted behind it.

A subtle effect of mounting an antenna on a metal tower is that sometimes the angular direction of a null will be changed so that it no longer provides maximum discrimination against a co-channel station.

Antenna Types

Several different types of antennas have been used in cable tv systems. One of the more popular types is called the *Yagi* after the man who invented it in the late 1920s. A Yagi antenna consists of three or more elements arranged as shown in Fig. 7-14. The element to which the transmission line is connected is called the *driven element*. It is basically a half-wave dipole, which if used alone would have a very broad pattern. Behind the driven element is a somewhat longer element that acts as a *reflector* to minimize signal pickup from the rear and to narrow the beam in front. One, or often many, shorter elements in front of the driven element, called *directors,* narrow the beam still further and increase the gain in the forward direction.

The only electrical connection between the transmission line and the Yagi antenna is to the driven element. Coupling to the other elements is accomplished electromagnetically. For this reason, the reflector and directors are called *parasitic elements,* and the Yagi is called a *parasitic array.* In general, a Yagi antenna will provide more forward gain for a given size and weight than any other type of antenna. It does, however, have many limitations. Both the pattern and the impedance at the terminals vary considerably with frequency. It is nearly impossible to hold the standing-wave ratio lower than about 2 to 1 over the bandwidth of a 6-MHz television channel. The front-to-back ratio will also vary as much as 5 dB or more over this frequency range. Yagis with many elements that are designed for maximum forward gain will have side lobes that are large enough to cause co-channel problems. Lastly, because of the sizeable back and side lobes, the pattern is quite easily affected by the mounting structure.

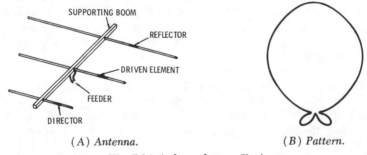

(A) *Antenna.*   (B) *Pattern.*

Fig. 7-14. A three-element Yagi antenna.

A newer type of antenna that is being widely used is the *log-periodic* antenna, an example of which is shown in Fig. 7-15. This is one of a class of "frequency-independent" antennas. These antennas are based on the discovery that if the elements of an antenna are made proportional to each other by some ratio, the antenna will have the same properties at frequencies related by the same ratio. The name log-periodic comes from the fact that lengths and spacings of the elements are periodic functions of the logarithm of frequency. Theoretically, there is no limit to the bandwidth of a log-periodic antenna. In practice, the low-frequency limit is dictated by the size of the whole structure, and the upper frequency is limited by the precision with which the shorter elements are constructed and positioned.

Practical log-periodic antennas have been made to cover many channels so that one antenna can be used to pick up signals from several stations in the same direction. The beam width and impedance vary somewhat with frequency, but these variations are well within tolerable limits.

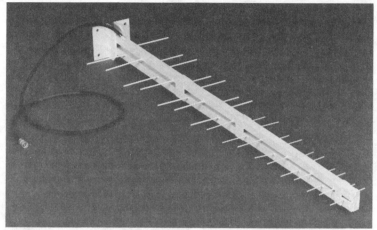

Fig. 7-15. A log-periodic antenna.

Several other types of antennas are widely used in cable tv systems. At the uhf channels where the wavelength is small, it is common to use a parabolic reflector driven by some element such as a dipole or a log-periodic antenna. In general, no one type of antenna is best suited for all situations. A typical cable tv headend may have several different types of antennas mounted on the same structure, as shown in Fig. 7-16.

Antenna Arrays

An antenna *array* is made up of several elements arranged to produce a pattern that is better suited to a particular purpose than the pattern of any one element used alone. There are two basic types of arrays. In the *end-fire* array, the elements are arranged one behind the other, and the major lobe of the pattern is along the axis of the elements. The Yagi and log-periodic antennas are actually end-fire arrays of elements. The other type of array is called a *broadside* array. The elements are placed side by side in a plane, and the major lobe of the pattern is at a right angle to the plane of the array. Most arrays are a combination of both types. Thus, it is common to find a broadside array whose elements are log-periodics or Yagis, which are in themselves end-fire arrays.

Fig. 7-16. A typical head-end antenna system.

In a broadside array, the array itself has a pattern that multiplies the pattern of the individual elements to form an overall pattern. The properties of the overall pattern can be changed by changing any or all of the following factors:

1. The number of elements
2. The spacing between elements
3. The magnitude of the coupling to each element
4. The phase of the coupling to each element

An understanding of the principles of antenna arrays can be gained by considering the two antennas shown in Fig. 7-17. For the sake of simplicity, the two elements in the figure are assumed to have isotropic patterns. It can be seen in Fig. 7-17B that as the spacing (S) between the elements is increased, symmetrical major lobes develop. These lobes become longer and narrower as the spacing is increased. After the spacing becomes greater than ½ wavelength, minor lobes develop that are oriented at a 90-degree angle with respect to the main lobes. The main lobes continue to increase and become narrower as the spacing is increased until a spacing of ⅝ wavelength is reached. At this spacing, the minor lobes are about 13 dB below the main lobes. If the spacing is increased further, the main lobes will decrease, and the minor lobes will increase until at one-wavelength spacing the lobes have equal magnitudes.

There is nothing magic about the way the pattern changes with element spacing. The field at any point can be found by vectorially adding the fields from the individual elements. The phase of the

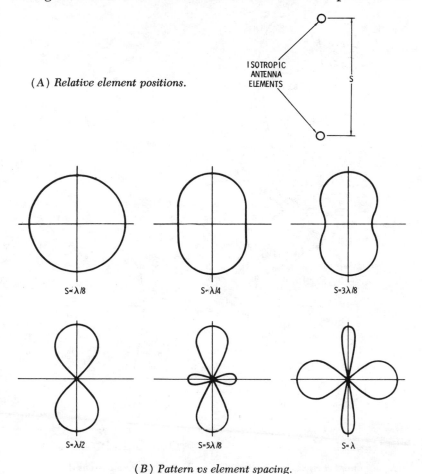

(A) *Relative element positions.*

(B) *Pattern vs element spacing.*

Fig. 7-17. **Effect of element spacing on the pattern of an array of antennas.**

fields at any point depends on their spacing; thus, at some points the fields will add, whereas at others they will cancel.

In addition to the spacing, the relative amplitude of the coupling to each element will also influence the shape of the pattern. Fig. 7-18 shows patterns of an array of five elements spaced ½ wavelength apart. To keep the illustration simple, only the forward half of the pattern is shown. This is not important because the back lobes can be controlled either by using a reflector or by using elements that have very little backward pickup. In Fig. 7-18A, all elements are in phase and are coupled equally to the transmission line. The pattern consists of a main lobe that has a half-power beam width of 23 degrees and side lobes that are about 13 dB down from

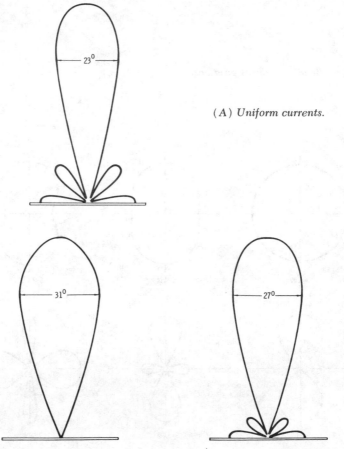

(A) Uniform currents.

(B) Binomial current distribution. (C) Tchebychev current distribution.

Fig. 7-18. Patterns of a five-element array for various current distributions.

the main lobe. There are four nulls in the pattern. This arrangement gives the highest gain and directivity that can be obtained from this type of array, but the side lobes are large.

If the phase of the connection to the transmission line is unchanged, but the amplitude of the coupling is altered by using directional-coupler–type devices, the side lobes can be reduced. When the contribution of each element is tapered from the center to the sides of the array, the side lobes will become lower, and the main lobes will become smaller and broader. When the contribution of each element varies as a so-called binomial series, the minor lobes will disappear completely. Under this condition, the main lobe has a half-power beam width of 31 degrees, as shown in Fig. 7-18B. For many applications, this is a desirable arrangement. In many applications, however, the main lobe is too broad to afford the necessary protection.

The narrowest beam width for any given level of minor lobes can be obtained by using a distribution that has been widely used in the design of electrical filters. It is called a *Tchebychev* polynomial distribution. An example is shown in Fig. 7-18C. This arrangement is considered optimum for an array design in many applications.

In the example of shape control shown in Fig. 7-18, only the amplitudes of the contributions of the various elements of an array were changed. It is also possible to change the pattern by changing the phase of the various elements. If the phase of each element is shifted progressively across the array, the entire pattern will be skewed to one side. This technique is useful in shifting a pattern in such a way that the desired station is still within the main lobe, but a co-channel station falls in one of the nulls of the pattern.

Antenna Specifications

The best overall picture of the behavior of an antenna or array is gained from a plot of its pattern. In addition to the pattern, many of the properties of an antenna are stated in the form of specifications. Among these are the following.

Minimum gain in main lobe—This is usually stated using an isotropic antenna as a reference. It may also be stated using a half-wave dipole as the reference. Remember that there is over 2-dB difference between the two references.

Side-lobe depression—This is the amount in decibels by which the side lobes are lower than the main lobe.

Front-to-back ratio—This figure is a measure of the back lobe of the antenna.

Horizontal beam width or half-power beam width—This is the angle in degrees between the directions at which the pickup of the antenna falls off by 3 dB.

Vertical beam width—This is the half-power width of the main lobe in the vertical plane.

Impedance and vswr—The impedance of a cable tv antenna is usually 75 ohms so that it can be connected directly to a 75-ohm cable. The degree of mismatch is usually stated in terms of vswr. This is the standing-wave ratio that would be measured on the transmission line if the antenna were used for transmitting. It is a measure of how well the antenna will match a transmission line.

Bandwidth—The bandwidth is a measure of the frequency range over which the above specifications will hold, or at least this is what it should mean. Care should be taken to see that the specifications are all within tolerable limits over the entire frequency range of interest. Some antennas have excellent specifications at their center frequency, but these degrade rapidly toward the edge of a 6-MHz television channel.

Mechanical specifications—These are straightforward and include size, weight, wind resistance, and weather resistance. In many latitudes, the effects of both wind loading and icing must be taken into consideration.

CHAPTER 8

The Headend—
Signal Processing

At the headend of the cable tv system, the signals from the antennas are processed for transmission along the coaxial cable distribution system. The functions that must be performed include one or more of the following:

1. *Amplification.* The signals picked up by the antennas are weak and must be amplified before they can be applied to the cable system.
2. *Adjustment of signal level.* The signals from the antennas are not all the same strength. The signals from local stations will be much stronger than those from more distant stations. In the cable, all signals that are carried should be very close to the same level.
3. *Automatic Gain Control.* Most received signals will vary considerably in level with time. The level of the signal applied to the cable system should be as constant as practical. In the signal-processing equipment, agc compensates for fading of the received signal.
4. *Conversion to another channel.* The highest frequency that the distribution cable system can handle is about 300 MHz, even in super-band systems. It is therefore necessary to convert all uhf signals to an unused vhf channel. Often, it is also necessary to convert local signals to a different channel because the same signal may be picked up off the cable and by the circuits of the receiver. The result is that the receiver will synchronize

137

on the stronger signal, which is usually from the cable, and will display the picture that is picked up directly as a ghost. The signal in the cable is delayed more than the off-the-air signal, so the ghost will lead rather than lag the main picture.

5. *Generation of Pilot Carriers.* Most modern systems use at least one, and often more, pilot carriers to provide a reference for automatic gain and slope control in the distribution system. These carriers must be generated at the headend.

6. *Combining of Signals.* In the signal-processing equipment, each channel is handled separately. In the distribution portion of the system, all channels are carried on the same cable. The last unit in the chain at the headend is a combiner that combines all of the signals to be carried onto one cable with a minimum amount of interaction between channels.

PROCESSING THE SIGNAL

All of the functions of the headend are usually performed by equipment that is combined into a rack and is called by the generic name, *signal processor.* Several different types of processors are in common use. A typical signal-processing setup is shown in Fig. 8-1.

Signal-to-Noise Ratio

It has been pointed out several times in this book that every amplifier in the system will introduce additional noise into the system. The signal-to-noise ratio is best at the headend and cannot be improved in the rest of the system, which merely adds more noise. The current FCC requirements call for a signal-to-noise ratio of at least 36 dB at the subscriber's terminal. Therefore, the signal-to-noise ratio at the headend must be much better than this.

The first step that can be taken to improve the signal-to-noise ratio is the use of an antenna with high gain. Fig. 8-2A shows a dipole connected to a signal processor through a length of coaxial cable. For practical purposes, we can consider all the noise to be referred to the input of the signal processor. The signal level at the output of the antenna is -10 dBmV. Since the loss of the cable is 6 dB, the signal level at the input of the processor is -16 dBmV. The equivalent noise input level at the processor is assumed to be -50 dBmV, so the signal-to-noise ratio at this point will be:

$$-16 \text{ dBmV} - (-50 \text{ dBmV}) = -16 + 50 = +34 \text{ dB}$$

which is too low to provide a good picture. If the antenna is replaced by one having a gain of 8 dB over a dipole, as shown in Fig. 8-2B, the signal level will be 8 dB higher at the input of the proces-

Courtesy National Cable Television Association

Fig. 8-1. A typical headend signal processor.

sor, and hence the signal-to-noise ratio will also be 8 dB higher, or +42 dB. It is thus seen that the signal-to-noise ratio in a system will be improved by the gain of the antenna. For this reason, it is usually advisable to use a high-gain antenna for all but the very strongest signals.

Preamplifiers

The signal-to-noise ratio of any system is usually controlled by the first amplifier in the system and can be improved by placing the first stage of amplification as close to the antenna as possible. The principle underlying this consideration is not easy to visualize, and the use of the decibel notation tends to obscure rather than clarify the situation. To simplify the situation, we will consider signal and noise powers in microwatts rather than in dBmV levels. To further simplify things, we will use absurd values of power, and thus avoid

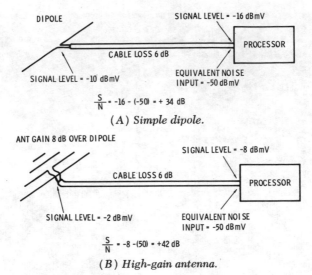

(A) Simple dipole.

(B) High-gain antenna.

Fig. 8-2. Examples of signal-to-noise ratio computations.

the use of decimals and fractions. The principle derived in this way will hold true when the noise powers have practical values.

Fig. 8-3A shows a signal of 10 μW connected to the input of a length of coaxial cable that has a loss of 6 dB. A loss of 6 dB means that the power at the output of the cable is one-quarter of the value at the input. The signal at the end of the cable is then 2.5

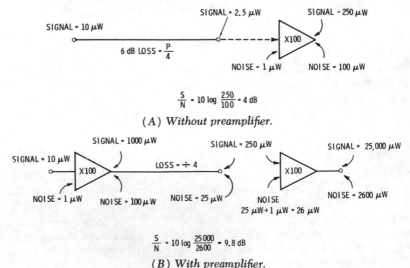

(A) Without preamplifier.

(B) With preamplifier.

Fig. 8-3. Comparison of signal-to-noise ratios.

140

μW. The equivalent noise-power input of the amplifier is assumed to be 1 μW. Both are amplified in an amplifier having a power gain of 100 times. The powers at the output are then 250 μW and 100 μW respectively. The signal-to-noise ratio in decibels is then

$$\frac{S}{N} = 10 \log (250/100) = 4 \text{ dB}$$

In Fig. 8-3B, an identical amplifier is placed directly at the output of the antenna before the cable. The signal power is again 10 μW and the noise power, 1 μW. At the output of the amplifier, both are amplified 100 times, and one-quarter of both appears at the end of the cable. The signal, which is 250 μW at this point, is simply amplified 100 times in the next amplifier. The noise, which has a power of 25 μW, is added to the 1-μW equivalent input noise power, and the sum is amplified 100 times. Thus, as shown in the figure, the signal power at the output is 25,000 μW, and the noise power is 2600 μW. The signal-to-noise ratio is then

$$\frac{S}{N} = 10 \log (25,000/2600) = 9.8 \text{ dB}$$

It is interesting to note that although two amplifiers will contribute more noise than one amplifier, the signal strength that would otherwise be lost in the cable more than makes up for the noise of the second amplifier as far as signal-to-noise ratio is concerned.

Another point of interest is that the signal-to-noise ratio at the input of the first amplifier is 10 dB, and it is nearly this value at the output of the second amplifier. The fact that the two figures are not exactly the same means that the second amplifier does contribute some noise to the system. In general, the contribution of the second stage to the noise performance decreases as the gain of the first stage is made higher.

The example of Fig. 8-3, in which we considered signal and noise powers in microwatts rather than in dBmV levels, is useful in understanding how noise powers add up in a system. It is, however, not very convenient for use in practical systems when we are trying to figure out how much a preamplifier will improve the noise performance of a given system. For this purpose, it is easier to work with noise factor. In Chapter 2, the noise factor, F, was defined as:

$$F = \frac{S_1/N_1}{S_2/N_2}$$

where,

S_1 is the input signal power,
S_2 is the output signal power,
N_1 is the input noise power,
N_2 is the output noise power.

Noise figure (F_N) is defined as

$$F_N = 10 \log \frac{S_1/N_1}{S_2/N_2}$$

where the symbols have the same meaning as in the preceding equation.

It can be seen that noise factor is a dimensionless number representing the ratio of two powers, whereas noise figure is this same ratio expressed in decibels. A perfect amplifier would contribute no noise at all and would have the same signal-to-noise ratio at its input and output. Its noise factor would then be 1, and its noise figure would be 0 dB.

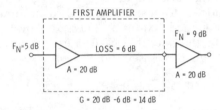

Fig. 8-4. First amplifier controls noise figure.

When two amplifiers are connected together through a cable, it is useful to consider the cable as being a part of the first amplifier, as shown in Fig. 8-4. Here the net gain of the first amplifier is equal to the gain of the amplifier in decibels less the loss of the cable, also in decibels. The noise factor, F, of the two amplifiers connected in tandem is given by

$$F = F_1 + \frac{(F_2 - 1)}{G_1}$$

where,
  F_1 is the noise factor of the first stage,
  F_2 is the noise factor of the second stage,
  G_1 is the net gain of the first stage, that is, the gain from the
    input of the amplifier to the output of the cable.

Thus, in Fig. 8-4, the first amplifier has a noise figure of 5 dB and the second stage of 9 dB. Both amplifiers have a gain of 20 dB, and the cable between them has a loss of 6 dB. We must first convert the noise figures to noise factors and the gain of the first amplifier (allowing for the loss of the cable) to a ratio, rather than a gain in decibels. Using the curves given in Chapter 2, we find that:

$$F_1 = \ 3.16$$
$$F_2 = \ 7.94$$
$$G_1 = 25.12$$

142

The noise factor of the two stages is then

$$F = 3.16 + \frac{(7.94 - 1)}{25.12} = 3.44$$

and the noise figure would be

$$F = 10 \log 3.44 = 5.36 \text{ dB}$$

This shows quite dramatically that although the second stage has a noise figure of 9 dB, the noise figure of the two stages in tandem is only 5.36 dB, compared to 5 dB for the first stage operating alone. Thus, the use of a preamplifier on the tower close to the antenna will improve the noise performance of any system in which the cables from the antenna to the signal processor are long.

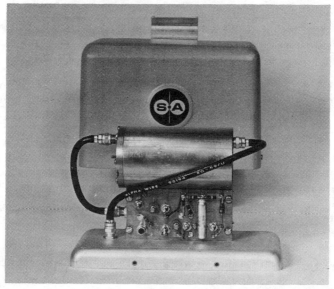

Courtesy Scientific-Atlanta, Inc.

Fig. 8-5. A cable tv preamplifier.

Most preamplifiers have input and output impedance of 75 ohms so that they can be used directly with regular coaxial cables without impedance-matching devices. A typical preamplifier is shown in Fig. 8-5.

It is important that the preamplifier be tuned to the particular channel to be amplified and have a low response to other frequencies. The reason for this is that any strong signal may drive the amplifier into nonlinear operation in which spurious signals will be generated. This is particularly apt to occur in locations

143

where the headend is in close proximity to a radio transmitting antenna.

Separating Signals

Unlike the amplifiers that are used along the cable, which handle all of the signals carried by the system, preamplifiers and signal processors in the headend handle only one signal in each channel. A separate channel is provided for each signal to be carried. The reason for this is obvious. It is much easier to design such a system than it is to design a broadband system. The signals that are picked up are at different levels and require different amounts of amplification, and broadband capability is not necessary at this point in the system.

When two or more broadcast stations are located in the same direction from the headend site, it is possible to receive them on the same antenna, if a broadband antenna such as a log-periodic is used. These signals must be separated before being fed to the signal processor or preamplifier, and the device that separates the signals must do so without mismatching the impedances or introducing ghosts. One device that is used for this purpose is called by such names as *hybrid, hybrid splitter,* or *hybrid ring.* There are many forms of hybrid splitters. All have the properties of the unit shown in Fig. 8-6, namely, if a signal is applied to the port marked 1, half of the signal will appear at port 2 and half at port 3 with no signal at port 4. Furthermore, any reflection into port 3 will not be seen at port 2 and vice versa.

There are many ways to make a unit that will function like the one shown in Fig. 8-6. The earliest such units were three-winding transformers called hybrid transformers, which were used by telephone companies to connect an amplifier into a line in such a way that the input and output circuits would be isolated. Perhaps the easiest way to understand how a hybrid operates is to consider the arrangement of transmission-line sections shown in Fig. 8-7. Here the lengths of the sections are chosen to control the amount of phase

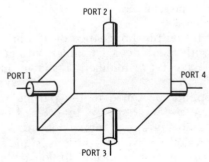

Fig. 8-6. Hybrid splitter.

shift in each path around the ring. First, assume that a signal is applied to port 1. The top path to port 4 has a total length of one-half wave, or a phase shift of 180 degrees. The path from port 1 to port 4 along the bottom ring has a phase shift of a full wavelength, or 360 degrees. Thus, the signals from the two paths arriving at port 4 will be exactly 180 degrees out of phase and will cancel each other. Now consider the paths from port 1 to port 2. The top path has a phase shift of 90 degrees. The bottom path has a phase shift of 360 + 90 degrees, which is the same as 90 degrees, so the two signals add at port 2. By the same reasoning, they also add at port 3.

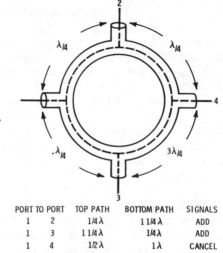

Fig. 8-7. Hybrid ring or "rat race."

PORT TO PORT		TOP PATH	BOTTOM PATH	SIGNALS
1	2	1/4 λ	1 1/4 λ	ADD
1	3	1 1/4 λ	1/4 λ	ADD
1	4	1/2 λ	1 λ	CANCEL

Now consider what would happen if a signal were introduced at port 2. Tracing the left and right paths around the ring, it can be seen that this signal would cancel at port 3. By the same token, a signal introduced at port 3 would cancel at port 2. This means that any reflection at port 2 or port 3 would not interact at the opposite port. This is exactly the arrangement needed to split the signal from one antenna into two separate paths that will not interact with each other.

It is interesting to note that a hybrid can also be used to combine two signals and couple them to one port. Hybrid devices are used as splitters and combiners in push-pull cable tv amplifiers.

A typical application in which a hybrid is used to split signals in a headend is shown in Fig. 8-8A. Here the signal from one antenna is split into two paths—one will handle one signal, and one will handle the other. Since only half of the signal will appear at the

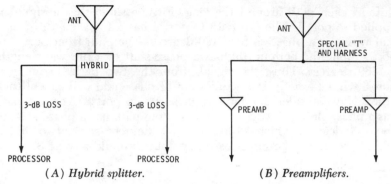

(A) Hybrid splitter.   (B) Preamplifiers.

Fig. 8-8. Two methods of splitting signals.

input of each processor, the signal level will be reduced by one half, or 3 dB. The signal-to-noise ratio will also be degraded by 3 dB. This may or may not be serious depending on the available signal strength of each channel.

It is possible to separate any two nonadjacent channels from a single antenna by using two preamplifiers with a coaxial T and a special cable harness. This arrangement is shown in Fig. 8-8B. The amplifiers are tuned to accept energy at only one channel frequency, so there is no signal loss. The harness and cable lengths are chosen to minimize reflections.

Bandpass Filters

It was pointed out above that one of the problems in the headend is that signals from nearby transmitters may be strong enough to drive a preamplifier or signal processor into nonlinear operation, even though the preamplifier or processor may be tuned to a completely different frequency. The result is the generation of spurious signals and, in extreme cases, intermodulation in which the modulation of one signal appears on the other.

To avoid overloading in strong signal environments, it is common to use a bandpass filter tuned to the channel of interest ahead of the preamplifiers. Fig. 8-9 shows the response curve of a bandpass filter tuned to Channel 9. Notice that the insertion loss is kept as low as possible in that channel, with the response nearly flat across the bandwidth of this one channel. The attenuation at the sound and visual carrier frequencies of adjacent channels is also shown.

Filters of the type used in this application differ from those used at lower frequencies. At low frequencies, filters have inductances and capacitances, but bandpass filters at cable frequencies are cavities that look like ordinary tin cans. It is not immediately obvious how an empty can will act as a resonant circuit. Fig. 8-10A

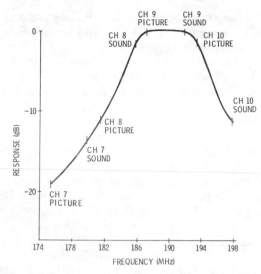

Fig. 8-9. Response curve of a typical bandpass filter.

shows a shorted section of a transmission line that is one-quarter of a wavelength long. A property of this type of line is that the reflected signal is in phase with the applied signal at the input terminals. This means that at the input terminals the line looks like an open circuit at the frequency at which its length is one-quarter wavelength. At other frequencies higher and lower than this, it looks like a capacitive or an inductive reactance. In this respect, it looks electrically like a parallel-resonant circuit. In Fig. 8-10B, several shorted quarter-wave sections are connected in parallel. Since they look like

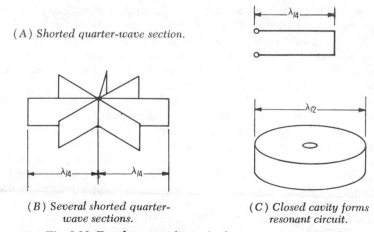

(A) Shorted quarter-wave section.

(B) Several shorted quarter-wave sections.

(C) Closed cavity forms resonant circuit.

Fig. 8-10. Development of a cavity from quarter-wave sections.

open circuits at the resonance frequency, it does not make any difference how many we connect in parallel. In Fig. 8-10C, we have connected all of the lines together and have a solid, walled cavity, which we can see will act like a parallel-resonant circuit. Energy is coupled into and out of the cavity either by short probes that protrude through the cavity or by small loops located inside.

Signal Converters

It is necessary to convert all uhf signals to unused vhf channels before applying them to the cable system. In some cases, this may be accomplished in the processor itself, or it may be accomplished by using a separate converter ahead of the processor. Converters for this purpose are packaged in weathertight cases, like preamplifiers, so that they can be mounted on the tower near the antenna. This will help the noise figure in the same way that mounting a preamplifier near the antenna will.

Converter design is particularly critical because a converter is inherently a nonlinear device, and stray signals that get into it will cause spurious signals.

Co-Channel Interference and Ghosts

It was pointed out in the preceding chapter that co-channel interference is often a serious problem in cable tv systems. The effect is minimized in home receivers by the separation of co-channel stations and by offsetting the carrier frequencies of co-channel stations that will put signals into the same geographical area. In cable tv, the antennas are mounted on tall towers to provide the best pickup of weak signals. This means that they will also pick up undesired signals. When the bearing to a co-channel station is many degrees away from that to a desired station, interference can be minimized by using antenna arrays that are phased so that they place a pattern minimum in the direction of the offending station. Unfortunately, at some headend locations the desired and offending stations are in the same general direction. Eliminating interference in such a case is nearly impossible at the antenna.

A relatively new device known by the trade name Phazar has been developed that will cancel co-channel interference whenever the desired and undesired signals can be picked up so that the level of one is higher on one antenna and the level of the other is higher on another antenna. A photograph of the device is shown in Fig. 8-11, and a block diagram is shown in Fig. 8-12. The system operates by combining the desired and undesired signals in one channel with the undesired signal in another channel. The level of the undesired signal is controlled by an attenuator, and its phase is controlled by a phase shifter. These are adjusted so that the undesired signal in the

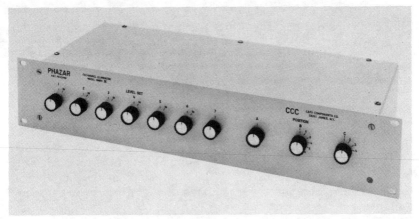

Fig. 8-11. Device for eliminating co-channel interference.

right channel of Fig. 8-12 is equal and opposite to the undesired signal in the left channel. The two versions of the undesired signal will cancel in the combiner, while the desired signal will be passed.

For proper operation, the two antennas should be mounted very rigidly so that their elements will not flex in the wind. This would introduce level and phase variations that would defeat the operation of the system. The antennas should be separated by at least ten feet, and equal lengths of feeder cable should be used.

This system can also be used to reduce other types of interference. In one case, the second harmonic of an fm transmitter located a few hundred feet from the headend produced a higher level at the visual carrier frequency of channel 10 than the desired signal. A small

Fig. 8-12. System for cancelling
co-channel interference.

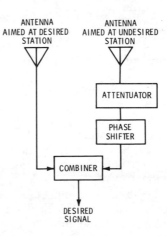

channel-10 Yagi antenna aimed toward the antenna of the fm station provided enough signal for 20-dB attenuation. In the same way, this system can be used to reduce ghosts. The undesired antenna is pointed toward the object responsible for the signal reflection that is causing the ghost.

SIGNAL PROCESSORS

The equipment that is actually used to modify the signals so that they are suitable for transmission along the cable system is called by the generic name of *signal processor*. This term is rather loosely used and usually includes all of the equipment in the building at the headend. There are three methods of signal processing in common use.

1. *Demodulator Processor.* This is the most elaborate form of signal processing. Each signal is demodulated to video and audio frequencies, called the *baseband*, before levels are adjusted. After processing, the signal is modulated onto a new carrier that often is not the same channel as the one on which it was received. This type of processor is called a mod-demod, or simply a demod processor.
2. *Heterodyne Processor.* In this type of processor, the incoming signals are heterodyned to an intermediate frequency, not the baseband, and the levels are adjusted. Then the signals are amplified and heterodyned to the channel that is to be transmitted along the cable.
3. *Strip Amplifier Processor.* This is the oldest and simplest form of cable tv signal processor. It consists essentially of a separate rf amplifier for each channel that is to be carried on the system.

Before describing the processors in detail, we will look at some signal and circuit considerations that, while not unique to cable tv, influence its design and adjustment.

The Phase-Locked Loop

The phase-locked loop is not new. Circuits operating on this principle were developed in the 1930s. However, in the days when vacuum tubes were the only active circuit components available, it was a difficult circuit to design and adjust. The advent of solid-state components designed specifically for this application has made it practical to use a phase-locked loop wherever it will lead to improved system performance.

Fig. 8-13 shows a block diagram of the basic phase-locked loop. It consists of three functional units: a phase detector, a voltage-controlled oscillator, and an amplifier-filter. The purpose of the system is to lock the phase, and hence the frequency, of an oscillator to the phase of a reference signal. The device may be used as both a signal generator and a detector.

The reference signal and the output of the voltage-controlled oscillator are applied to a phase detector. The output of the phase detector is proportional to the phase difference between the two signals. This signal is amplified and filtered and applied back to the voltage-controlled oscillator to bring its phase into coincidence with that of the reference signal. Assume that the frequency of the voltage-controlled oscillator is the same as that of the reference signal. The output of the phase detector will then be a dc voltage that is proportional to the phase difference between the two input signals. The polarity of this voltage is such that it will change the phase of the signal from the voltage-controlled oscillator in the proper direction to obtain phase lock. At this point the two oscillators are in synchronism. Actually there will be no output from the phase detector when the two signals are in phase. When the oscillator tends to drift out of synchronism, a dc voltage will be provided by the phase detector. Only a very small signal is necessary from the phase detector because the signal is amplified many times in the amplifier-filter stage.

Note that this is a feedback system, and if the constants were not of the correct value, it could oscillate or hunt. Under this condition, the phase and frequency of the voltage-controlled oscillator would move to try to get to the correct value, but they would overshoot and oscillate back and forth around the correct point. This is prevented by the filter in the amplifier-filter stage which adjusts the response so that the system will be stable.

When the system is first turned on, the frequency of the signal from the voltage-controlled oscillator will not be the same as the frequency of the reference signal. The process by which the frequency of the voltage-controlled oscillator is brought to the correct value is quite complicated. In practice, the two frequencies can be

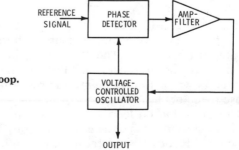

Fig. 8-13. Phase-locked loop.

pulled to the same value if their difference is not greater than the bandwidth of the amplifier filter. Fig. 8-14 shows the frequencies of the reference signal and the signal from the voltage-controlled oscillator. The range over which the frequency of the voltage-controlled oscillator can be brought to the reference frequency is called the *pull-in range*. Once the oscillator is locked, it will stay locked over a wider range. That is, the frequency of the reference signal may be changed considerably before the system falls out of lock. This range, labeled in Fig. 8-14, is called the *hold-in range*.

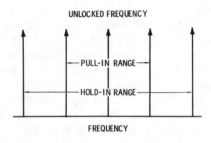

Fig. 8-14. The range of a phase-locked loop.

The most obvious use of the phase-locked loop is to generate a signal that has exactly the same frequency as another signal. The reference signal might, for example, be supplied by a very stable crystal-controlled oscillator. The phase-locked loop may also be used to generate several precise frequencies from a single crystal

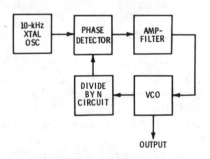

Fig. 8-15. A frequency synthesizer using a phase-locked loop.

oscillator. Fig. 8-15 shows a circuit in which a single 10-kHz crystal oscillator is used to generate many different frequencies that are separated by 10 kHz. The 10-kHz signal is used as the reference signal, but the frequency of the voltage-controlled oscillator is divided down to 10 kHz before the resulting signal is applied to the phase detector.

The phase-locked loop may be used in many different ways in a cable tv system. One is as a synchronous detector in the headend, which will be described later in this chapter.

Quadrature Distortion

Another factor that influences the design and adjustment of the signal processor is that the vestigial-sideband transmission system used for television signals introduces a unique form of distortion known as *quadrature distortion*. This can be understood by first considering an rf carrier that is modulated by conventional double-sideband a-m. We will assume that both the carrier and modulating signals are simple sine-wave signals. The spectrum of the modulated signal will consist of the carrier and two sidebands. The sidebands will be separated from the carrier by the frequency of the modulating signal, as shown in Fig. 8-16A. All three of these signals may be represented by rotating phasors. The lowest frequency, f_L, will rotate the slowest; the carrier frequency, f_C, will rotate faster than f_L; and the highest frequency, f_H, will rotate the fastest of the three.

To simplify things, we can consider the carrier phasor to be stationary and the two other phasors to rotate in opposite directions about the tip. The resultant modulated signal is the sum of the three phasors. When the two modulation phasors are above the horizontal, they will increase the value of the resultant, as in Fig. 8-16B. When they are below the horizontal, they will decrease the length of the

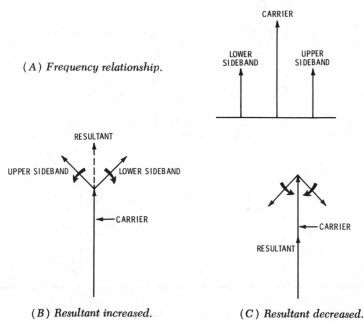

(A) *Frequency relationship.*

(B) *Resultant increased.*   (C) *Resultant decreased.*

Fig. 8-16. Phasor representation of a double-sideband a-m system.

resultant phasor, as shown in Fig. 8-16C. The resultant phasor is then a vertical line that becomes longer and shorter with modulation and represents the envelope of the modulated wave. Since the rotating phasors are always on opposite sides of the carrier phasor, the resultant will always be vertical. This is the same as saying that the modulation changes the amplitude of the resultant wave but does not change its phase.

Now let us consider a single-sideband signal. Here we have the carrier phasor and only one sideband phasor which rotates about its tip, as shown in Fig. 8-17A. The resultant is still the sum of the two phasors and will become longer and shorter with modulation, but now the resultant will shift back and forth as shown in Figs. 8-17A and 8-17B because there is only one modulation phasor and it is on either one side or the other of the carrier phasor. Thus, there will be some phase shift of the carrier with modulation in addition to the amplitude modulation.

This phase shifting of the resultant can be thought of as the result of one component in phase with the carrier being added to

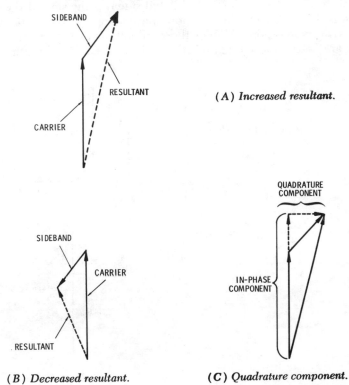

(A) *Increased resultant.*

(B) *Decreased resultant.*   (C) *Quadrature component.*

Fig. 8-17. Phasor representation of ssb signal.

another component that is 90 degrees out of phase, or *in quadrature*. For this reason, the resulting distortion is called quadrature distortion.

The television signal is not strictly a single-sideband signal. At modulation frequencies up to about 750 kHz, it is a combination of regular a-m and single sideband. Above this frequency, for all practical purposes, it is single-sideband modulation.

It will be seen later that the type of detector used in a headend demodulator will determine the seriousness of quadrature distortion.

Demodulator Processors

Fig. 8-18 shows a block diagram of a demodulator used in a demodulator processor. The first stage is a mixer that heterodynes the received signal with a signal from a local oscillator to produce an i-f just as in a regular superheterodyne receiver. The local oscillator is crystal controlled to ensure stability.

The mixer may be of any type, but balanced mixers using two diodes are becoming increasingly popular because they cancel second harmonics in the same way that a push-pull amplifier does. Separate i-f amplifiers are provided for the aural and visual signals. Each has its own agc system. This separation of the picture and sound signals permits separate adjustment of levels.

The output from the video i-f amplifier is fed to a demodulator, which may be either an envelope detector or a synchronous demodulator. The relative advantages of the synchronous detector will be discussed later. A delay equalizer is included to compensate for the phase shift that is inevitably introduced by the bandpass amplifiers in the system. The video signal is then amplified.

The sound portion of the signal is heterodyned to 4.5 MHz so that it will be easy to apply it to the modulator later. Both audio and video signals are usually available for monitoring.

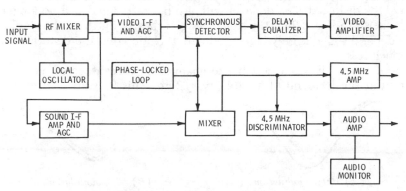

Fig. 8-18. Block diagram of a demodulator.

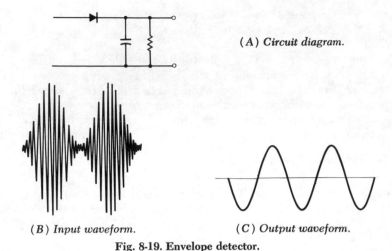

(A) Circuit diagram.

(B) Input waveform.

(C) Output waveform.

Fig. 8-19. Envelope detector.

The Envelope Detector

The envelope detector is actually a half-wave rectifier just like the type used in some power supplies. Fig. 8-19 shows an envelope detector, together with the input and output waveforms. If the input is a double-sideband a-m signal, the output will be a faithful reproduction of the modulating signal. This is shown in Fig. 8-19C. If, on the other hand, the input is a single-sideband signal, the output will be distorted as a result of the quadrature distortion discussed earlier.

Fig. 8-20A shows the output of an envelope detector when the input is a 100%-modulated single-sideband signal. The output for a double-sideband signal is also shown. It can be seen that the lower peaks of the envelope are sharpened, which means that the signal will be distorted. The dc level will also be shifted and will no longer be zero. In Fig. 8-20B, the same thing is shown, but here the single-sideband signal is only 50% modulated. Notice that the distortion is much less.

Of course, it is not possible to change the percentage of modulation at the headend of a cable tv system. This is controlled at the

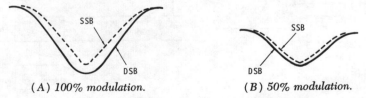

(A) 100% modulation.

(B) 50% modulation.

Fig. 8-20. Waveforms from envelope detector.

broadcast transmitter. What can be done at the headend, however, is to tune the detector stage in such a way that the level of the sideband is reduced as compared with the carrier. This will have the same effect as reducing the percentage of modulation and will reduce the quadrature distortion.

The Synchronous Demodulator

The way that the synchronous demodulator works can be appreciated by looking back for a moment to the envelope detector shown in Fig. 8-19A. It can be seen that the diode conducts when the signal has one polarity and is shut off when the signal has the reverse polarity. This same effect could be produced with the circuit of Fig. 8-21A if we had some way of closing the switch when the signal had one polarity and opening the switch when the signal had the reverse polarity. The output would be exactly the same as that from the envelope detector.

One form of synchronous demodulator is shown in Fig. 8-21B. The diode bridge is turned on and off by the reference signal. If this reference signal is phase locked to the carrier frequency, the output will be the modulation envelope.

The synchronous demodulator has the advantage that its switching can be controlled by simply shifting the phase of the carrier signal that is applied to the bridge. In this way, the switching can be timed so that the demodulator will ignore the quadrature component of the signal that leads to quadrature distortion.

The synchronous demodulator is also better from the standpoint of delay distortion and is finding increasing use in cable tv headends.

The Modulator

After the signals are demodulated at the headend, they must be remodulated onto another carrier. This is accomplished in a modulator. Fig. 8-22 shows a block diagram of a typical modulator. The

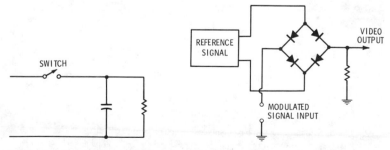

(A) *Simplified envelope detector.*   (B) *Synchronous-detector circuit.*

Fig. 8-21. Synchronous demodulator.

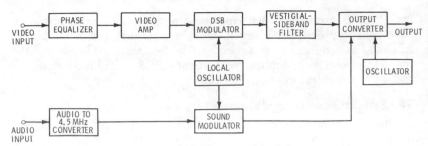

Fig. 8-22. Block diagram of modulator.

inputs are the video signal and either an audio signal or a 4.5-MHz audio carrier. The video signal is modulated onto a carrier in an ordinary double-sideband modulator. The signal is then passed through a vestigial-sideband filter that produces a signal with the same spectrum as one from a broadcast station.

In some systems the carriers are all produced by a system that phase-locks them to a master 6-MHz oscillator. This tends to minimize the effects of distortion from cross modulation and intermodulation.

The demodulator type of processor is the easiest to design and adjust because it provides separate control of each function. Inasmuch as the input to the video section of the modulator is a video signal, this type of processor is very convenient in systems that originate programming. The video from the camera chain can be applied directly to the modulator input. Fig. 8-23 shows a portion of a demodulator processor using modular construction so that units can be added as needed.

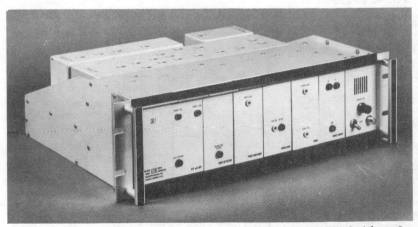

Courtesy Scientific-Atlanta, Inc.

Fig. 8-23. Portion of typical demodulator processor.

Heterodyne Processors

In the heterodyne processor, the incoming signal is heterodyned to an intermediate frequency and then is amplified and heterodyned back to one of the vhf, midband, or superband channels. A block diagram is shown in Fig. 8-24, and a photo of a typical unit is

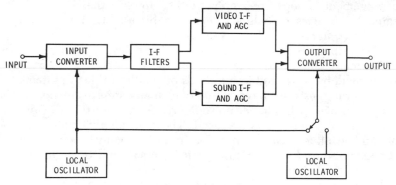

Fig. 8-24. A heterodyne processor.

shown in Fig. 8-25. The input and output mixers use the same local oscillator when the signal is to be transmitted on the cable at the same frequency as that on which it was originally broadcast. When it is desired to change channels, two separate local oscillators are used. Most heterodyne processors have one or more substitute carrier oscillators so that locally originated programs can be fed to the cable.

Courtesy Scientific-Atlanta, Inc.

Fig. 8-25. A typical heterodyne processor.

In the processor shown in Fig. 8-24, the video and audio i-f's are separated to provide better control of the levels. Some processors use the same i-f amplifier for both audio and video i-f's. In this case, control of carrier level is accomplished by a tuned trap that reduces the level of the audio carrier.

159

The heterodyne processor does not provide the flexibility of the demodulator processor, but it is more economical. A well-adjusted heterodyne processor will provide excellent-quality signals.

Strip-Amplifier Processors

The *strip-amplifier* processor is just what its name implies—an arrangement of separate amplifiers that amplify the signals on the same frequencies on which they were broadcast. It is sometimes called a *straight-through* processor.

The strip processor is the earliest and simplest type of processor. It has many limitations as far as use in large multichannel systems is concerned, but it is used in hundreds of smaller cable tv systems.

Modern transistors have resulted in many improvements to the strip amplifier. It has lower noise figures than other types of processors, unless preamplifiers are used. The strip amplifier also has the advantage of not using mixers or frequency converters; therefore, it does not introduce as many spurious signals and is not as subject to intermodulation.

Early systems of this type had very poor agc characteristics and provided very poor control of carrier levels. A modern strip amplifier will provide just as much agc as a heterodyne processor and will provide good control of carrier level with tuned traps. Although it has many limitations, the strip amplifier is often the only economically viable processor for small systems in rural areas. If it is not pushed beyond its inherent limitations, it is capable of providing high-grade signals.

Many smaller systems use a combination of processors. In these installations, strip amplifiers are used for signals to be carried on the same channel, and some form of converter or heterodyne processor is used for the other channels.

Generation of Pilot Carriers

Many cable tv amplifiers use one or more pilot carriers for agc and asc along the cable. These carriers must be generated at the headend of the system so that they will experience the same conditions as the signals being carried by the system.

Pilot carriers are generated either by separate crystal-controlled oscillators or by oscillators that are phase locked to a crystal standard.

Combining Signals

Regardless of the type of processor used, the last unit at the headend must combine all of the signals for transmission along one coaxial-cable distribution system. In addition to the television signals picked up off the air, there are often locally originated signals and possibly some fm channels. Some systems carry all of the signals in use on the fm band at the frequencies on which they were originally broadcast.

There are two methods that are commonly used for combining signals. In strip amplifiers, the output circuits are carefully tuned so that they can be connected in parallel without excessive interaction between channels. This arrangement has the advantage of not introducing appreciable loss; therefore, it cuts down on the amount of amplification required in the headend. Although the signal-to-noise ratio is established primarily by the first stage of amplification, the amount of distortion will increase as more amplifiers or stages of amplification are used.

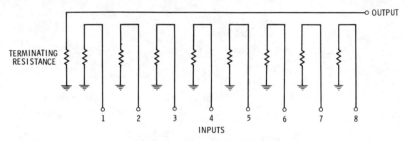

Fig. 8-26. An eight-channel combiner.

The more common method of combining signals in a way that will not allow interaction between channels is to use a device, appropriately called a combiner, that is similar to several directional couplers connected in series. Fig. 8-26 shows a diagram of an eight-channel combiner. The circuit has an input impedance of 75 ohms at each of its ports. It acts just the opposite of a directional coupler; signals introduced at the ports are coupled to the output but not to each other, and reflected signals are dissipated in the 75-ohm termination.

CHAPTER 9

Powering the Cable TV System

Because a cable tv system is spread over a wide area, there are unique problems in supplying operating power to its various components. The amplifiers of a cable system are often spaced several thousand feet from each other and often cover an entire community. Each amplifier must be supplied with operating power.

In many of the older systems that use vacuum-tube amplifiers, there is often a connection to the power line at each amplifier location. This is an expensive arrangement, and as a result these systems were designed with a spacing between amplifiers that is much greater than the optimum spacing. With the advent of newer amplifiers that use much less operating power, it became practical to transmit the operating power along the same coaxial cable that carries the signal.

Because transistor amplifiers operate on low dc voltages, it would seem ideal to transmit a low dc voltage along the cable with the tv signal. This is impractical because of the corrosion resulting from electrolysis.

A cable tv system uses several different types of metals, such as copper and aluminum. Whenever two dissimilar metals are connected together in the presence of moisture and the normal impurities that are found in the atmosphere, the junction tends to act like a small battery, and electrolysis will result. When a direct current flows through a series of such junctions, it will aid some of the small "batteries" and oppose others. As a result, the electrolysis, and hence the corrosion, will be increased considerably at some of the junctions. This is why connection at the positive pole of an automobile battery will corrode more than the negative terminal. To avoid this effect, almost all present-day cable tv systems use ac operating power transmitted along the coaxial cable. The use of dc operating power is not dead, however, and a great deal of development work is being done to make it practical.

COUPLING AC POWER TO THE COAXIAL CABLE

There are three considerations involved in feeding 60-Hz operating power to a coaxial cable that also carries tv signals:

1. The signal in the cable must not be shorted out through the power supply.
2. Noise and possibly interfering signals that might be present on the power line must not be coupled into the cable system.
3. The configuration of the coupling device must not cause reflections on the cable.

The device that actually couples the 60-Hz power onto the cable is called a *power inserter*. It is constructed in much the same way as a signal splitter or directional coupler and can be mounted on a pole or suspended directly from the cable. The power inserter (Fig. 9-1) contains filtering arrangements that will allow the power to flow into the cable but will block the signal from flowing into the power supply.

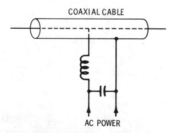

Fig. 9-1. Power inserter.

The construction of the power inserter is very important. The relationship between all of the conductors must be such as to make the coupler look just like a section of 75-ohm cable to the television signals.

The electrical performance of a power inserter is described in terms of the return loss at its input and output signal connectors and in decibels of isolation between the cable and the power source as far as the signal is concerned.

THE AC POWER SUPPLY

In any power transmission system, the losses are proportional to the square of the current flowing in the conductors. Thus, the most economical transmission system would use the highest practical voltage because less current would be required to transmit a given amount of power. From this point of view, it would be desirable to use the highest possible voltage for transmitting power along the cable. The coaxial cable itself will withstand quite high voltages, but safety considerations and local ordinances governing the pole lines that are used impose some restrictions. The highest voltage normally used is 60 volts, and in localities where this is not allowed, 30 volts is used. Most present-day amplifiers are designed to operate from either a 30-volt or a 60-volt supply.

The ac voltage applied to the cable should be regulated so that the high and low voltage limits of the individual amplifiers will not be exceeded. This is particularly important when power connections are made at the extreme ends of a power distribution system where the line voltage may vary over a wide range.

Voltage regulation is accomplished by deriving the voltage from a voltage-regulating, or constant-voltage, transformer connected between the power line and the power inserter. Several different configurations are used in voltage-regulating transformers, but the principle can be understood from the circuit of Fig. 9-2.

In the circuit shown in Fig. 9-2, the primaries of two transformers are connected in series, and their secondaries are connected so that the secondary voltages will oppose each other. Transformer T1 is an ordinary transformer. Transformer T2 has a saturable core and is normally operated with the core at least partially saturated. When the line voltage increases, the primary voltages of both transformers will increase, but the core of the lower transformer will saturate more, changing the division between the primary voltages in such a way as to keep the secondary voltage constant. For convenience,

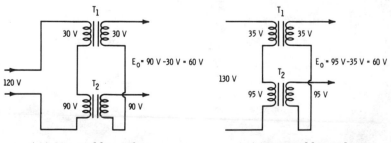

(A) *Nominal line voltage.*   (B) *Increased line voltage.*

Fig. 9-2. Constant-voltage transformer.

both transformers in Fig. 9-2 have a simple 1-to-1 ratio. In Fig. 9-2A, the line voltage is 120 volts, and it divides with 30 volts across the primary of T1 and 90 volts across the primary of T2. The output voltage is the difference between the two secondary voltages, or 60 volts. In Fig. 9-2B, the line voltage has increased to 130 volts. Because the core of the lower transformer is saturable, the division of voltage between the two transformers changes. Now we have 35 volts across the top transformer and 95 volts across the primary of the lower transformer. The output voltage, which is the difference between the two secondary voltages, is still 60 volts. Transformer arrangements of this type will provide a nearly constant secondary voltage over a wide range of line-voltage variations. Several different arrangements have been used to make transformers of this type. One of the more popular is a three-legged transformer in which one of the legs is saturated.

All power supplies should be mounted in well shielded containers like that shown in Fig. 9-3. Shielding of the power supply is important because it is a place where interfering signals might get into the cable.

Cable tv power supplies also include devices for protecting the system from lightning surges. This subject is covered in more detail in Chapter 19.

The voltage-regulating transformer used in the cable power supply does not produce a sinusoidal secondary voltage but produces a flattened waveform as shown in Fig. 9-4. This is actually an advantage because with a flattened waveform more power can be transmitted for a given peak voltage, and the peak voltage is what

Fig. 9-3. A cable tv power supply.

Courtesy AEL Communications Corp.

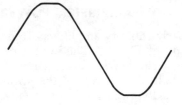

Fig. 9-4. A typical waveform from a
constant-voltage transformer.

is limited by the safety considerations. The unusual waveform can be a bit of a problem when the voltage is to be measured.

The amount of power transmitted is determined by the rms value of the voltage. This is the value that we specify in a power supply. Most ac voltmeters are calibrated to indicate rms voltage, but this calibration is only valid when the voltage being measured is sinusoidal. Fig. 9-5A shows a sinusoidal voltage. The relationship among the peak, average, and rms values is given. An ac voltmeter that uses a D'Arsonval mechanism, such as an ordinary vom, actually responds to the average value of the applied waveform. The scale is marked so that the indication will be 1.11 times the average value of the measured voltage. When the voltage is sinusoidal, this indication is the rms value.

Fig. 9-5B shows a square wave of voltage. Here, the average, peak, and rms values are all equal. Thus, if an ordinary vom, which is calibrated to measure sinusoidal voltages, were used to measure the value of a square-wave voltage, the indication would be 11% high.

The waveform of a cable tv power supply is not a perfect square wave because the resulting harmonics would be very difficult to filter out. As a result, there is no simple correction factor that can be applied when a D'Arsonval type of meter is used. The iron-vane type of ac voltmeter responds to the square of the voltage being measured and is calibrated in terms of rms voltage. On a square or nearly square wave, it is more accurate than the D'Arsonval type. As a result, iron-vane meters, which are rarely used in other electronic circuits, are frequently used to measure the operating voltage on a cable tv system.

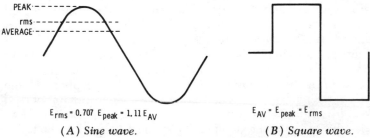

$E_{rms} = 0.707\ E_{peak} = 1.11\ E_{AV}$ $E_{AV} = E_{peak} = E_{rms}$

(A) Sine wave.   (B) Square wave.

Fig. 9-5. Relationship between rms and average values of voltage.

AMPLIFIER POWER-SUPPLY CIRCUITS

Early cable tv amplifiers often used the simple half-wave rectifier circuit of Fig. 9-6. The output of this circuit can be filtered to provide a ripple-free direct current, but since current flows only on one half of the cycle of the applied voltage, the current in the line is actually a pulsating direct current and not a true alternating current. A pulsating direct current is just as bad as far as corrosion is concerned as a pure direct current. To overcome this, some of the amplifiers on a system used the circuit of Fig. 9-6A in which current flows only during the positive half of the cycle, and others used the circuit of Fig. 9-6B in which current flows only during the negative half cycle. If the amplifiers on the system were balanced, the current on the cable would flow during both half cycles.

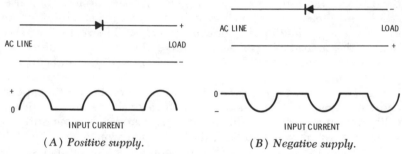

(A) *Positive supply.*   (B) *Negative supply.*

Fig. 9-6. Half-wave rectifier circuits.

This arrangement has been abandoned in favor of the full-wave bridge rectifier shown in Fig. 9-7. This circuit is easier to understand when it is drawn as shown in Fig. 9-7A. Many schematic diagrams show it in the form given in Fig. 9-7B. Actually, the two circuits are exactly the same.

The principal limitation of the bridge rectifier circuit is that there is a diode between the input and output grounds, so these grounds are not at the same potential. This is evident in Fig. 9-7B. Since the case of the amplifier and the outer shield of the coaxial cable should be at the same potential, it is customary to use isolation transformers in amplifiers that use bridge rectifiers.

Another rectifier circuit that operates on both halves of the input voltage wave, but also has a common ground between the input and output circuits, is shown in Fig. 9-8. This circuit is actually a voltage doubler. On the half of the input voltage cycle when the top wire is negative, current flows through diode D1, charging capacitor C1 to the peak value of the input voltage. On the next half cycle, diode D2 conducts, charging capacitor C2. When the output current is low, the

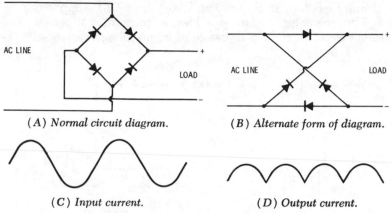

(A) Normal circuit diagram.

(B) Alternate form of diagram.

(C) Input current.

(D) Output current.

Fig. 9-7. Bridge rectifier circuits and waveforms.

voltage that charges capacitor C2 is equal to the sum of the line voltage and the voltage across capacitor C1, so the output will be twice the peak value of the applied voltage. With a normal load, the voltage is somewhat lower than this value. The voltage regulation of a voltage doubler is rather poor, so it is usually followed by a voltage regulator.

As we will see in the following section, there is a voltage drop along the coaxial cable due to its resistance. As a result, the voltage will be reduced considerably at the amplifier that is the greatest distance from the power supply. Voltage-regulator circuits are used to permit amplifiers to operate at different distances from the point where power is applied to the cable. Most cable tv amplifiers can operate satisfactorily over a wide range of applied voltage. The usual range of operating voltages is 18 to 30 volts for an amplifier rated at 30 volts, and 36 to 60 volts for an amplifier rated at 60 volts.

Fig. 9-9 shows a simplified schematic diagram of a power-supply circuit. Inductor L1 and capacitor C1 form a low-pass filter circuit that prevents the tv signal from getting into the power supply. The switching arrangement permits the power transformer to be connected to either the input connector of the amplifier or the output connector, or both. Thus, the amplifier can get its operating power from the cable at either end, or the power can pass through the amplifier to other amplifiers farther along the cable. Usually the

Fig. 9-8. Voltage-doubler circuit.

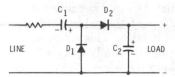

transformer has taps so that either a 30-V or a 60-V supply can be used. The regulator circuit provides a constant voltage to the operating circuits over a wide range of input-voltage variation.

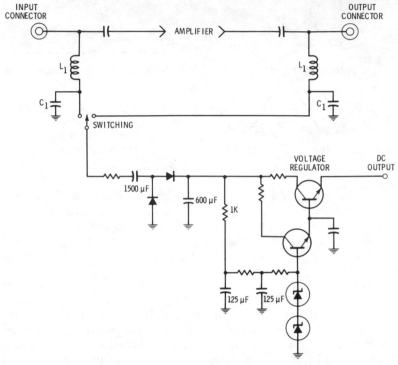

Fig. 9-9. Typical amplifier power supply.

CABLE RESISTANCE AND VOLTAGE DROPS

In transmitting any voltage along a coaxial cable, there will be losses due to the resistance of the cable. That is why we have to use amplifiers. Just as the signal is attenuated when passing through the cable, the ac power-supply voltage will suffer a voltage drop also. The effect of cable losses on the power-supply voltage is much less than on the signal. At the 60-Hz power frequency, the skin effect is not serious, and the voltage drop is caused primarily by the resistance of the cable itself. Most of the resistance is in the inner conductor simply because it is smaller, but it is customary to specify the effect of the cable on the power-supply voltage in terms of its *loop resistance*. This is the equivalent resistance of a given length of cable and is usually specified in terms of ohms per 1000 feet of cable. The loop resistance ranges from a fraction of an ohm per 1000 feet for one-inch–diameter cable to over 50 ohms per 1000 feet for the solid dielectric cable that is sometimes used for distribution and drop cables.

To determine the voltage drop that can be expected between amplifiers, we must know the length of the cable, its loop resistance, and the current that it is carrying. The design is carried out so that the amplifier that is the farthest removed from the power inserter will always have its minimum required operating voltage.

Fig. 9-10 shows a small section of a cable tv system. The cable used in this part of the system has a loop resistance of 0.5 ohm per 1000 feet, and an attenuation at channel 13 of 0.72 dB per 100 feet. The symbols used in Fig. 9-10 are explained in Appendix A. In this particular system, the amplifiers are spaced 19 dB apart. Remembering that the spacing is based on the loss at channel 13, we can calculate the distance between the amplifiers in feet as follows:

$$\text{Distance} = \frac{19}{0.72} \times 100 = 2639 \text{ ft}$$

Knowing that the physical length of the cable between amplifiers is 2639 feet and the loop resistance is 0.5 ohm per 1000 feet, we can easily find the total resistance, R, of the cable between amplifiers:

$$R = \frac{2639 \text{ ft}}{1000 \text{ ft}} \times 0.5 \text{ ohm} = 1.32 \text{ ohms}$$

This resistance of 1.32 ohms will be the same between all of the trunk amplifiers shown in the figure. The voltage drops will not all be the same, however, because the current is different in different sections of the cable.

Looking at Fig. 9-10, we see that the current between points A and B is 0.5 A. Thus, the voltage drop across this section is

$$E = 1.32 \,\Omega \times 0.5 \text{ A} = 0.66 \text{ V}$$

Since the power-supply voltage is 60 V, the voltage at the input to the amplifier at point B will be 59.34 V. The drop across this par-

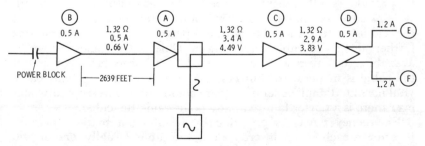

Fig. 9-10. Computation of voltage drops in cable.

ticular section of cable is so small as to be insignificant. The current for the amplifier located at point A does not flow through the cable, so it does not contribute to the voltage drop.

To find the voltage drops in the sections of cable to the right of point A in the figure, we must first find the current in each section. Starting with the points farthest away from point A, we find that the currents for lines E and F together with the current drawn by the amplifier at D total

$$1.2\ A + 1.2\ A + 0.5\ A = 2.9\ A$$

This is the current in the section of cable between points C and D. The current between points A and C is 0.5 A higher because of the current drawn by the amplifier at point C. Thus, the drop between A and C is

$$3.4\ A \times 1.32\ \Omega = 4.49\ V$$

and between C and D the voltage drop is

$$2.9 \times 1.32\ \Omega = 3.83\ V$$

The voltage at point C is then 55.5 V and at point D, 51.7 V. Along the lines at E and F there will be additional voltage drops which will depend on the lengths of E and F.

It is worth noting that the power-blocking arrangement, which is shown in Fig. 9-10 as a capacitor, may not actually be a capacitor. All this symbol means is that there is some provision at this point that prevents the ac power from passing.

EMERGENCY POWER

In the early days of cable tv when most systems were small, emergency power was not considered to be very important. Usually when there was a power failure, all the tv sets on the system were without power, and there was little to be gained by keeping the system operating. Furthermore, the principal use of the system was to provide entertainment and not vital services. Many modern systems cover entire cities, and it is altogether possible that part of the cable system may be without power while customers' sets farther along the cable have power. Unless emergency power is available, customers will be frustrated if their sets have power but the cable is inoperative. Furthermore, many modern systems carry vital nonentertainment services that should not be interrupted whenever there is a power failure at any point along the cable.

Emergency power is provided by installing a battery-operated inverter at each power inserter along the cable. Usually, one or two 12-V automobile batteries are used to provide power to the inverter, which provides a 60-Hz voltage whose waveform is similar to that of a constant-voltage transformer. A provision is made for keeping the batteries charged from the ac line through a battery charger. To be effective, the emergency supply must automatically provide power to the system whenever the ac power fails. Usually, this transfer to the emergency supply is made in a matter of a few milliseconds and is not even noticed by the subscriber.

It is very helpful for troubleshooting if an external light is provided to indicate when the emergency supply is in operation. It is important that the emergency power supply have the same current-supplying capacity as the regular supply; otherwise, the voltages at the amplifiers on the cable will not be the same as in normal operation, and the performance of the system may be degraded.

HEADEND POWER

Providing power to the headend of a system is much more straightforward than providing power to the amplifiers along the cable. All of the components at the headend are located close together, and most of them operate from a 120-volt, 60-Hz line. The biggest problem usually is keeping noise and extraneous signals that might be present on the power line out of the system. There are many stages in a headend that operate at very low levels and are quite susceptible to noise and interference. It is usually advisable to install a filter in the power line at the point where it enters the headend building. Power-line grounding can also be a problem at the headend. Although one side of the line is grounded, the ground is often not at the same potential as the system ground. This can cause problems in three ways:

1. Lightning surges can get into the system. (See Chapter 19.)
2. Through improper grounds, noise and interference can get into the cable system.
3. Ground loops often introduce hum that will modulate the video and audio carriers.

The best approach is to install an rf filter at the point where the power line enters the building and connect the ground of the filter to the ground of the other headend equipment.

In tracing ground loops, it is helpful to remember that the path a signal follows is due to the skin effect on conductors. Fig. 9-11 shows a conductor entering a shielded enclosure through a hole in the shielding. Note that the path of any rf current will be along the outside of the conductors and the shielding because of the skin effect. The current will not flow through the metal shielding. Thus, in the figure the current flows in through the hole in the enclosure,

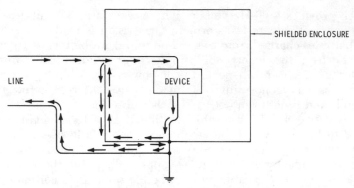

Fig. 9-11. Rf current path due to skin effect.

through the device inside the enclosure, around the inner surface of the enclosure, and back out through the same hole by which it entered. In general, an rf current will leave an enclosure through the same hole by which it entered. Remembering this simple principle can save hours of trouble in finding a ground loop or an ineffective shield.

Headend components that are usually mounted on a tower close to an antenna, such as preamplifiers and converters, are usually powered through the coaxial cable, just like the amplifiers that are distributed along the system. The principles are exactly the same as those described above in connection with supplying power to amplifiers along the cable. The loop resistance of the antenna feeder cables must be taken into consideration when you are finding the voltage drops. In general, the fedeers are short enough that voltage drop is not a serious problem.

FIELD POWER SUPPLIES

Whenever trouble in a system is isolated to a particular amplifier, it is often desirable to run a few tests on the amplifier before pulling it out of the system and taking it back to the repair shop. Unfortunately, most of the equipment that might be used to test an amplifier requires 120-volt ac power, and the only power available at most amplifier locations is the 30-volt or 60-volt ac power that is transmitted along the cable. The fact that the same pole might also carry regular power lines is of no value because there is no way to connect to them. A useful device for such situations is a field power supply that will operate from 30-volt or 60-volt power and supply 120 volts for operating other equipment. A typical unit of this type is shown in Fig. 9-12. This unit will plug into an amplifier and will supply 120 volts at an outlet on the front panel.

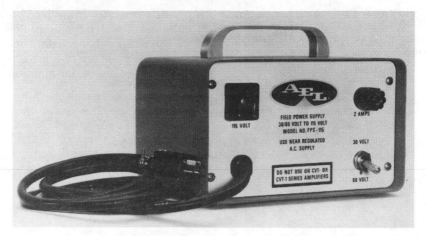

Fig. 9-12. Typical field power supply.

BENCH POWER SUPPLIES

When an amplifier is tested in the shop, it is important that the conditions be the same as when the amplifier is in service on the cable. This means that power should be supplied through the coaxial cable connector by means of a power inserter. A frequent cause of poor amplifier performance is that the amplifier was aligned in the shop under conditions that were far different from what it was exposed to in actual service. Even short leads may change the circuit parameters enough to disturb the alignment of the amplifier.

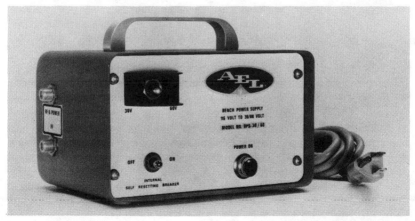

Fig. 9-13. Typical bench power supply.

Fig. 9-13 shows a bench power supply that operates from the 120-volt ac line and supplies either 30 volts or 60 volts for amplifier testing. Power is supplied by a power inserter through a coaxial connector. The amplifier input or output is connected to one coaxial connector on the power supply, and an instrument or termination device is connected to the other connector.

CHAPTER 10

Program Origination

The full potential of any cable tv system will not be realized until the system is capable of originating its own programming. The more a system has to offer, the more attractive it will be to subscribers. For this reason, almost all systems have some form of program origination. In some smaller systems, this is often nothing more than a small monochrome camera that continuously scans a clock and a few weather instruments with an occasional announcement poster. In larger systems, complete studios such as that shown in Fig. 10-1 are available for originating programs.

The reason that program origination has been so slow in spreading throughout the cable tv industry is that equipment capable of generating broadcast-standard television programs is very expensive. Only the larger systems can afford it. More economically priced cameras and recorders are available for industrial use, but in the past these have not been capable of producing broadcast-quality pictures.

Two separate tv standards have been issued by the Electronic Industry Association (EIA) to cover television signals. Standard RS-170 covers broadcast-quality signals, and standard RS-330, which is somewhat less stringent, covers industrial-quality signals. Until recently, most cable tv systems complied with the less-stringent standard for program origination.

Two technical developments have given an impetus to cable tv program origination. First, the rapid development of digital integrated circuits has made the development of high-quality synchronizing signals economical. Secondly, the advent of what is called *electronic journalism* or *electronic news gathering* has led to the development of a large number of cameras and recorders that are economical but are capable of producing broadcast-quality pictures. The device that did the most to make this practical is the *time-base corrector,* which will be discussed later in this chapter.

The equipment required to originate programs includes cameras, synchronizing-signal generators, and recorders. A complete treatment of all of these devices is beyond the scope of this book. However, this chapter does cover the basic principles and will serve as a basis for further study.

Courtesy National Cable Television Association

Fig. 10-1. A cable tv studio.

THE CAMERA

The first link in the chain of components that originate a television picture is the camera. Cameras used for origination of cable tv programs usually use either a vidicon or a lead-oxide vidicon tube. We will consider each of these tubes in some detail.

Camera Tubes

Fig. 10-2 shows a sketch of a *vidicon* tube. The inside surface of the glass faceplate is electrically conductive. On this conducting film

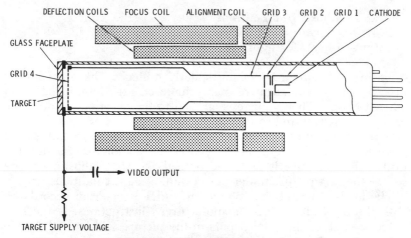

DEFLECTION COILS   FOCUS COIL   ALIGNMENT COIL   GRID 3   GRID 2   GRID 1   CATHODE

GLASS FACEPLATE

GRID 4

TARGET

VIDEO OUTPUT

TARGET SUPPLY VOLTAGE

Fig. 10-2. Diagram of vidicon tube.

is deposited a very thin layer of photoconductive material. When the tube is dark, each element of the photoconductive layer is an insulater. When an element of the layer is illuminated, it becomes slightly conductive, and the potential on the inside rises toward the potential of the signal electrode. The assembly looks electrically like a large number of tiny capacitors that are somewhat leaky. The amount of charge on the inside of any of the elements depends on the amount of light reaching it through the faceplate. Therefore, on the photoconductive layer there is a series of electric charges that have the same pattern as the light reaching the faceplate.

The other electrodes are provided to change this charge pattern into an electrical signal. The electron gun is similar to the electron gun in a tv picture tube. Grid 3 is a beam-forming electrode, and grid 4 provides a uniform field between itself and the photoconductive layer. The electron beam is scanned across the photoconductive layer in the same way that the beam is scanned in a picture tube.

When the electron beam scans the photoconductive layer, it deposits electrons until the layer reaches the potential of the cathode of the electron gun. Any excess electrons are bounced off and are not used. This depositing of electrons causes a current to flow from the target electrode. The current is a video signal that corresponds to the brightness of the scene being viewed.

Alignment of the electron beam in the vidicon is accomplished by a magnetic field from a focusing coil. The beam is deflected by deflection coils. Several of the electrode voltages are adjustable and are used to set the camera for optimum performance. In general, the effect of the voltage adjustments is:

179

Grid 1—Adjusts beam current
Grid 3—Adjusts picture focus
Target—Adjusts for light level

Setting the camera is not particularly difficult. The procedure is first to increase the negative bias voltage on grid 1 until the beam current is zero. Then voltage is applied to the other electrodes. The camera iris is adjusted so that minimum light reaches the tube. The bias voltage on grid 1 is then set so that the highlight details in the picture are just brought out as viewed on a monitor. Finally, the grid-3 voltage, the iris, and the optical focusing adjustments are set for the best picture under the prevailing light conditions.

The final setting of the voltage on grid 1 is made in accordance with the particular camera manufacturer's instructions. If the bias is too low, all of the bright spots in the picture will have the same brightness, and detail will be lost. If the bias is too high, the scanning spot will become too large, and picture resolution will suffer.

The vidicon provides excellent pictures and is highly compatible with the characteristics of the average tv picture tube. It does tend to "hang up" ("stick"), or retain its image, when adjusted for optimum picture quality at low light levels. It finds its widest application in film-pickup service where the light level can be kept reasonably high.

Another type of camera tube that is widely used in cable tv is the *lead-oxide vidicon*. This type of tube is often referred to by the name *Plumbicon,* which is a registered trademark of N. V. Philips of the Netherlands. (The name "Plumbicon" is derived from the Latin word *plumbum,* which means lead. Part of the target assembly of a Plumbicon is made of lead monoxide.) Fig. 10-3 shows a partial sketch of a lead-oxide tube. The electron gun and focusing electrodes are the same as in a conventional vidicon. The lead-oxide tube differs chiefly in the faceplate and target assembly. The inside of the glass faceplate is coated with a thin, transparent layer of tin

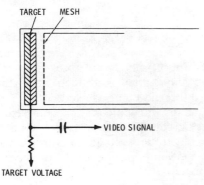

TARGET   MESH

VIDEO SIGNAL

TARGET VOLTAGE

Fig. 10-3. Lead-oxide vidicon.

oxide. This is the layer from which the output signal is taken. On top of this layer there is a layer of photoconductive lead monoxide.

The operation of the tube can be illustrated by the equivalent circuit shown in Fig. 10-4. Each element of the target looks electrically like a capacitor connected in parallel with a variable resistance. The scanning electron beam can be thought of as the moving pole of a switch that scans all of the capacitors and connects them to ground.

If the screen is dark, each of the capacitors charges to about 30 volts, the same voltage that is applied to the target through the signal electrode. Under this condition, there will be no output as the electron beam scans the target. When a section of the target is illuminated, the variable resistor that is in parallel with the capacitor will decrease in resistance. As a result, the capacitor will charge to a voltage that is lower than the supply voltage. When the electron beam scans this section of the target, enough current will flow to charge the capacitor to 30 volts. Thus, the output signal will increase as the light on each picture element increases.

The lead-oxide tube has many advantages over the conventional vidicon. Inasmuch as the dark current is very low, the tube has a low noise level and good performance at low light levels. It has much less tendency to "stick" and is less sensitive to heat than the vidicon.

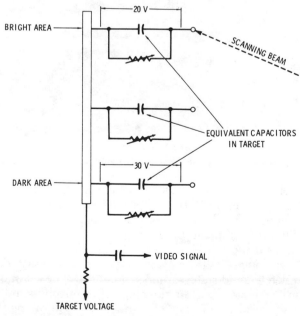

Fig. 10-4. Principle of lead-oxide vidicon.

A newer type of camera tube, called the *silicon-target vidicon,* is similar to the conventional vidicon except that the target is an array of extremely small silicon photodiodes. In operation, the diodes are reverse-biased when a positive voltage is applied to the n-type substrate of the target. Each diode acts like a tiny capacitor. When light strikes an area of the target, the corresponding capacitors are discharged. The more light, the more they are discharged. The scanning electron beam recharges each of the capacitors and provides a target current that is a function of the scene brightness. The

Courtesy International Video Corp.

Fig. 10-5. A color tv camera.

silicon-target vidicon has a very high light sensitivity, high resolution, and low dark current. Its performance represents a significant improvement over the older vidicons. Fig. 10-5 shows a color camera that uses Plumbicons in the blue and green channels and a silicon-target vidicon in the red channel.

A new development called a *charge-coupled device* may eventually make completely solid-state cameras practical. The charge-coupled light sensor consists of a single chip of semiconductor material covered with an array of tiny electrodes. It requires neither high voltage, a scanning beam, nor a vacuum envelope, but it does

Fig. 10-6. Experimental camera using charge-coupled device.

require a rather elaborate electronic circuit to scan the scene. Fig. 10-6 shows an experimental charge-coupled camera.

Lenses

The first element of a program-origination system that influences picture quality is the camera lens. In general, any degradation of the picture that is caused by the lens cannot be removed later in the system. It is essential, therefore, that the lens be of high quality and be kept clean at all times.

The basic principles of lenses are illustrated in Fig. 10-7. There are two basic types of lenses—the positive, or *convergent, lens* of Fig. 10-7A and the negative, or *diverging lens* of Fig. 10-7B. The converging lens focuses parallel light rays that strike one side to a point at the other side. The distance from the lens to the point is called the *focal length* of the lens. For most practical purposes. the focal length of a lens may be thought of as the distance from the lens to a screen where a distant scene is brought into sharp focus. The focal length of a lens is usually expressed in inches or millimeters.

A diverging lens (Fig. 10-7B) spreads parallel light rays and causes them to appear to diverge from a point in front of the lens.

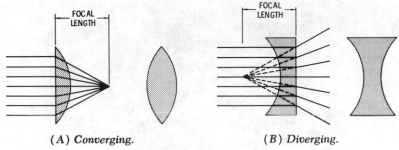

(A) *Converging.*   (B) *Diverging.*

Fig. 10-7. Types of lenses.

Again, the distance from the lens to the point is called the focal length of the lens.

Practical camera lenses are not simple converging or diverging lenses, but are compound lenses made up of several component lenses. The reason for this is that one component lens will compensate for some of the defects of another component lens, giving a compound lens of high quality. Fig. 10-8 shows a sketch of a typical camera lens.

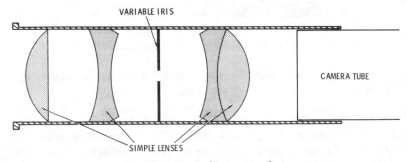

Fig. 10-8. Typical tv camera lens.

The distortions of a picture caused by a lens are called *aberrations.* Some of the more common aberrations are:

1. *Spherical aberration.* This results in picture distortion when the iris is wide open for viewing scenes at low levels.
2. *Chromatic aberration.* The focal length is different for different colors; as a result, some colors are slightly out of focus.
3. *Barrel or pincushion distortion.* Straight lines near the edge of the picture become curved. A pattern of straight lines will be shaped like a barrel or a pincushion, hence the name.
4. *Astigmatism.* Objects at the edges of the picture will be out of focus and will lack definition.

184

5. *Ghost images.* Reflection of light from the surfaces of the lenses will cause ghost images in the picture.

The focal length of the camera lens will determine the *viewing angle*, or the *width of field*. Most tv cameras have lenses of different focal lengths. A lens with a focal length of about 35 mm has a very wide angle of view and could probably pick up a scene the entire width of a studio. A longer-focal-length lens, called a telephoto lens, has a viewing angle of only about 3°. Cameras are usually equipped with a turret that has lenses of different focal lengths so that rapid changes between wide-angle shots and close-ups are possible.

A newer type of lens that is widely used has a variable focal length. This lens, called a *zoom* lens, permits changing the focal length by means of a mechanical adjustment while keeping the scene in focus.

A property of lenses that often is not well understood is called the *speed* of the lens. The speed of a lens is a measure of the amount of light that passes through the lens to the camera target. More light passes through a high-speed lens than through a low-speed lens, other things being equal. The formula for lens speed is:

$$f = F/D$$

where,

f is the lens speed,
F is the focal length of the lens,
D is the inside diameter of the lens.

Naturally, F and D must be expressed in the same units, so f is simply a number without units. Speed is usually expressed as an "f-number" such as $f2$ or $f3.5$. This is sometimes called an "f-stop."

Smaller f-numbers represent faster lenses. Looking at the formula, we see that the greater the diameter of a lens, or the shorter its focal length, the more light will pass. The f-number of a lens is usually inscribed on its case. The advantage of a higher lens speed is that the lens will pass more light and will provide better pictures under low light conditions.

The amount of light reaching the target of a camera is controlled by an iris (similar to the iris in an ordinary snapshot camera). The iris controls the size of an opening by means of leaves that slide together to reduce the opening. Inasmuch as closing the iris reduces the amount of light passing through the lens assembly, it will effectively increase the f-number of the assembly.

Another interesting property of a lens and iris assembly is called the *depth of field.* When a lens is focused on an object a given distance from the camera—say 20 feet—the object will be in sharp focus. Other objects that are slightly nearer to or farther from the camera will also be nearly in focus. That is, they will appear sharp enough for all practical purposes. Objects that are much closer to or farther from the camera will be out of focus. The range of distances over which an object will remain in sharp focus is called the depth of field. The greater the f-number of the lens and the smaller the iris, the greater the depth of field. Thus, "stopping down" the iris of a lens will increase the depth of field, but at the expense of reducing the amount of light that will reach the target of the camera.

Usually, studio lighting is maintained at high enough levels that a satisfactory depth of field can be obtained. Under extremely low light levels, care must be taken to keep objects of interest sharply focused, whereas in sunlit scenes almost everything that the camera can see will be in reasonably sharp focus.

Viewfinder

An important accessory of a television camera is the *viewfinder,* which is actually a miniature monitor mounted on the camera so that the operator can see it. Some earlier cameras used optical viewfinders, but these are rarely used today. The viewfinder enables the operator to observe the field and quality of the image that is being picked up by his camera.

The viewfinder uses a black-and-white picture tube, even though it might be on a color camera. It has the usual controls that are used with picture tubes, including brightness, contrast, and focus. In addition, controls are usually provided for horizontal and vertical size and linearity. By means of the size controls, the operator can see all of the picture, even the portion that is normally blocked out on the subscriber's receiver.

An important feature of some viewfinders is a provision for displaying pictures other than those picked by the associated camera. This is very helpful when two pictures are being superimposed. Each operator can see the composite scene and make any necessary adjustments that would affect his portion of the composite picture.

THE ENCODER

So far, we have discussed mostly cameras with a single camera tube. Some cable tv systems originate only black-and-white telecasts and therefore are interested only in monochrome cameras. Most larger systems, however, originate color programs.

The principle of operation of a color camera is similar to that of a monochrome camera except that three tubes are used to pick up the red, green, and blue portions of the scene. As explained in Chapter 3, these three color images are all that is required to reconstruct the complete scene. Some color cameras use a fourth tube to pick up the luminance portion of the scene. This has the advantage of making the luminance signal independent of the matrixing system.

Matrixing of the outputs of the three camera tubes is accomplished in what is normally called the *encoder*. The encoder is used simply to generate the composite color signal with both luminance and chrominance information. The technique used is similar to that described in Chapter 3. In fact, all of the equipment used in program origination is used to create a composite television video signal, so most of the material in Chapter 3 applies directly to program origination.

VIDEO TAPE RECORDING

Early television recording was accomplished on motion-picture film. Film projectors with built-in cameras are still widely used to show motion pictures on television. This equipment is large and expensive and certainly out of the reach of the small-system operator. Since its introduction in 1956, video magnetic tape has been replacing film in television recording.

The process of recording a video signal on magnetic tape is rather complex. The reason is that the bandwidth of a signal that can be recorded on magnetic tape is not great enough to accommodate a television video signal. This obstacle is overcome by first using the video signal to frequency-modulate an rf carrier, and then recording the modulated carrier on the magnetic tape. The resulting bandwidth may be as small as 1 MHz.

To anyone who has worked with communications systems, this seems impossible. It is considered axiomatic that the bandwidth of any communications system must be at least as great as the highest frequency of the signal. This is not necessarily true with fm recording. In frequency modulation, the *frequency* shift of the modulated carrier is proportional to the *amplitude* of the modulating signal. The *frequency* of the modulating signal determines the *rate* at which the fm carrier is shifted. In conventional fm broadcasting, the deviation ratio is described as:

$$\text{Deviation Ratio} = \frac{\triangle F}{f}$$

where,

$\triangle F$ is the maximum carrier deviation at 100% modulation,
f is the maximum modulating frequency.

The greater the deviation ratio, the greater the bandwidth of the modulated signal. By reducing the *amplitude* of the video modulating signal, we can reduce the deviation ratio and thus the bandwidth

of the modulated fm signal. In video recording, deviation ratios of
0.5 and under are used with a resulting bandwidth that is less than
that of the original video signal.

This compression of bandwidth looks as if we are getting something for nothing, but there is a penalty. As the bandwidth is reduced, the signal-to-noise ratio also decreases. The precision needed
to keep the signal-to-noise ratio within acceptable limits is one of the
reasons for the high cost of video tape recorders.

Quadruplex Recorders

There are two basic electromechanical techniques used for
actually recording the fm signal on magnetic tape. The higher-priced,
broadcast-quality recorders are called *quadruplex* recorders because

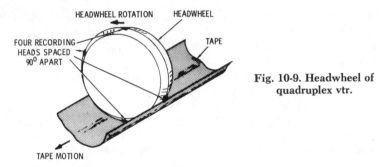

Fig. 10-9. Headwheel of
quadruplex vtr.

they use four recording heads. These heads are mounted on a *headwheel* as shown in Fig. 10-9. The heads scan across the tape so that
the recording consists of paths that are almost at right angles to the
direction of tape travel. As one head reaches the edge of the tape,
the next head starts recording on the next path, as shown in Fig.
10-10. The recording heads are spaced as close to 90 angular degrees
apart as is practical. In spite of this, there may be some error due to

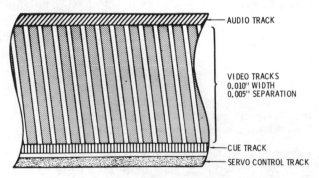

Fig. 10-10. Track format for quadruplex vtr.

the heads not arriving at the edge of the tape at the same relative time. This type of error is called *quadrature error* or *switching error* because it results when the recording is switched from one head to another.

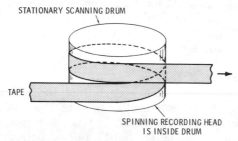

STATIONARY SCANNING DRUM

TAPE

SPINNING RECORDING HEAD
IS INSIDE DRUM

Fig. 10-11. Tape path in helical-scan vtr.

The speed of rotation of the headwheel and the speed of the tape are both very closely controlled. Usually a servo system is provided to regulate the speed automatically.

Helical-Scan Recorders

A lower-cost video tape recorder (vtr) called a *helical-scan* video recorder gets its name from the way in which the tape passes around a scanning drum. This is shown in Fig. 10-11. There is a slot in the scanning drum through which the pole pieces of one or two recording heads protrude. The recording head spins inside the drum. Because of the helical motion of the tape, the path of the recording is as shown in Fig. 10-12. Each scan covers one field of the video signal.

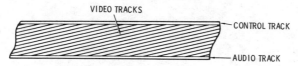

VIDEO TRACKS

CONTROL TRACK

AUDIO TRACK

Fig. 10-12. Typical track format for helical-scan vtr.

An inspection of Fig. 10-12 will show that the helical recording is much more sensitive to variations in tape speed than the quadruplex recording. The result is that a type of error called *time-base error* is introduced. Without some form of correction, a helical recorder usually produces a signal that is not up to standard broadcast quality.

Fig. 10-13 shows a helical-scan vtr of the type widely used for cable tv work. This vtr uses a version of fm called *pulse-interval modulation*, which results in improved performance. When used with a time-base corrector, its performance is entirely satisfactory and compares favorably with that of broadcast vtr's.

Synchronization

Some vtr's are equipped for synchronization with external signals. Such a recorder can be locked to the sync signals used in the studio in such a way that there will be no loss of sync during switching between the vtr and a local camera.

Dropout Compensation

One of the problems with vtr's is that the signal occasionally "drops out" due to defects in the tape or the recording process. Sometimes the tape momentarily lifts off the recording head. A degree of dropout elimination can be obtained from a circuit that may be either built into the recorder or contained in a separate unit. The circuit contains two paths by which the video signal may pass between the input and the output. One path is direct. The other path is through a delay line that stores one horizontal line of the

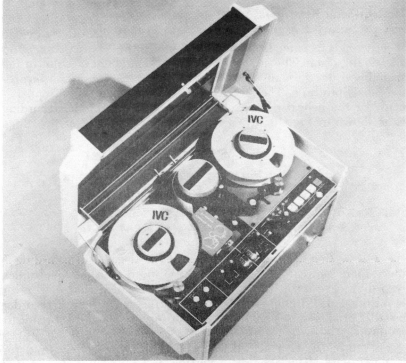

Courtesy International Video Corp.

Fig. 10-13. A helical-scan vtr.

190

picture. **Normally, the** signal passes through the unit without delay. **If a dropout is detected,** the signal from the delay line is switched **to the output. Thus, when a** dropout occurs, video information from **the preceding line of the picture** is sent to the output. This information is not exactly the same as that from the missing line of the picture, but it is usually sufficiently similar that an occasional brief dropout will not be noticed by the viewer.

Handling Video Tape

Many of the problems that are frequently encountered with video recorders can be eliminated, or at least minimized, by careful selection and handling of the video tape. The requirements imposed on magnetic tape used for recording video signals are stringent. During a half-hour recording, about 1/5 mile of tape is used, and the head of the vtr travels about 30 miles at a speed of about 60 mph.

Magnetic tape consists of a base of plastic tape that is coated on one side with magnetic material that is suspended in some sort of binder. The binder holds the particles of magnetic material in position and serves as a lubricant. Until recently, ferric oxide was the most commonly used magnetic material. Modern tapes use more elaborate crystalline cobalt or chromium magnetic materials.

The base of the tape is either acetate or polyester. Polyester tapes are stronger and have greater long-time stability. Some tapes are prestretched to avoid dimensional changes with time.

It pays to use high-quality recording tape. There are enough problems in video recording without introducing more by using tapes of poor quality. The tapes designed for use with helical recorders are not the same as those designed for use with quad machines. Although the particles of magnetic material are very small, they are longer in one dimension than in the other. The orientation of the particles is controlled so that they will magnetize best along the path of the machine with which they are designed to operate.

In handling magnetic tape, cleanliness is one of the most important considerations. A tape should be good for 100 or more passes through a vtr. If it does not last this long, the trouble is usually in the handling and storing of the tape. Particles of dust and dirt will shorten the useful life of a tape considerably. The storage area should be free from dust and smoke, and the tape should be stored in a closed container.

Tape should be stored vertically to avoid edge damage and should be protected from extremes of temperature. Tape may be cleaned with a solvent recommended by the manufacturer. Other solvents might dissolve the binder or the tape base. Loosely wound tapes may be wound firmly but not tightly.

Time-Base Errors and Correction

The most common faults in a picture recorded on a vtr are called *time-base errors*. These errors are caused by variations in the speed of the tape or the recording head. Other sources of time-base errors include variations in signal phase that occur in some micro-wave links, and variations in signals that are received through a satellite link. A slow vertical oscillation of a satellite with respect to the surface of the earth will introduce a slow varying time-base error in the signal.

Time-base errors include jitter, drift, picture tearing, "flag waving," hue shift, color streaking, and skew errors. In a vtr, there are three factors that contribute to time-base errors. First, there are defects in the tape itself; the tape may stretch or suffer geometrical changes due to the way it is stored and handled. Second, the tape transport mechanism in the vtr may introduce errors due to variations in tape speed or tape tension. Finally, changes in the speed of the recording head can introduce errors. Those vtr's in which the tape and head speeds are servo-controlled introduce less error than those in which the speeds are controlled by battery voltage or line frequency.

The video time-base corrector (tbc) has done more than any other device to make local program origination practical in cable tv systems. There is a great deal of confusion about what a tbc can and cannot do to a video signal. What it can do is remove time-base errors that were introduced during the recording or playback process or by the signal processing. It will increase resolution and will restore the gray scale and color to a replica of the signal that was originally applied to the vtr. What it will not do is add something to the signal that was never there in the first place. The linearity, color fidelity, and signal-to-noise ratio of a signal will never be better than the original signal that was recorded.

The tbc operates by varying the propagation time of the signal path. This is just the opposite of the effects that introduced the time-base errors into the signal. The earliest tbc's were built into expensive broadcast-quality vtr's. Now, many self-contained tbc's are available. They fall into two categories: those that operate on an analog principle and those that operate digitally.

Fig. 10-14 shows a block diagram of an analog tbc. At the input of the tbc, the signal is split into two paths. One path is through a delay line. The other is to a comparator. In the comparator, the time references of the signal are either measured in comparison with the studio sync signal, or measured to detect variations in the vertical, horizontal, or color sync of the signal. While the errors are being detected, the signal is stored in the delay line. By the time

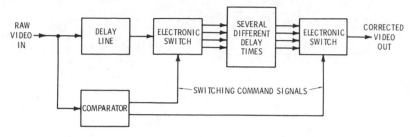

Fig. 10-14. Analog time-base corrector.

the signal has propagated through the delay line, the comparator has decided how much time should be added or subtracted from the propagation time through the system to remove the time-base errors from the signal. The time is varied by switching in the required amount of delay.

Fig. 10-15 shows a block diagram of a digital tbc. Here the input video signal is converted into a digital signal by an analog-to-digital converter. The digital signal is stored in a random-access digital memory. The system provides separate write and read clock signals so that the read rate can be adjusted to remove any time-base errors. The digital signal is then converted back to analog form.

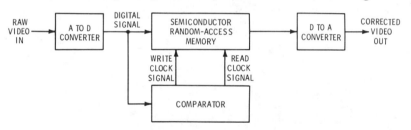

Fig. 10-15. Digital time-base corrector.

There are a wide variety of tbc's comercially available, and not all of them will work with all types of vtr's. The capability of a tbc is often expressed in terms of a time *window*. If a tbc can compensate for errors over a range of 180 microseconds, it is said to have a 180-microsecond window.

Serious time-base errors that are outside the range of a tbc can sometimes be eliminated by recording from a helical vtr to a quad vtr through a tbc, and then playing back through a tbc onto the cable.

CHAPTER 11

The Complete System

So far we have discussed the components of a cable tv system independently of each other. This informaiton will apply to most systems because as far as the types of components are concerned, one cable tv system looks very much like another. They all use antennas, headends, cables, and amplifiers. Here the similarity ceases. There are probably no two complete cable tv systems that are exactly alike.

This dissimilarity between systems is responsible for more troubles than any other factor. The technician lacks a standard of comparison for his complete system. When dealing with a component, the technician can test it, adjust it, and if necessary repair it, until it meets the manufacturer's specifications. At that point he will be reasonably confident that the amplifier will behave as well as can be logically expected. With a complete system the story is different. A system may perform better or worse than some other system, but the two systems are usually so different that the technician can never be certain that he is getting optimum performance.

In a newer system that has been designed entirely in one step, there is usually a good set of specifications for performance at various stages of the system. These can serve as benchmarks for evaluating repairs and adjustments. More often, however, a system has grown like Topsy, with new channels added as the signals became available and new feeder lines added as the area developed. In such cases there is rarely a good set of system specifications.

The problem of system specifications becomes more serious when it is necessary to perform a major modification to a system, such as adding two-way capability. If the performance standards of the original system are uncertain, its performance after the modification can merely be speculated on.

It is possible to reconstruct a set of specifications for a system that is already in existence and, in so doing, to isolate the weak links in the system. These can either be corrected or factored into any future plans for expansion.

SYSTEM PHILOSOPHY

A cable system must be thought of as a link in a chain that starts with the scene being televised and that ends on the screen of the subscriber's tv set. The adage that a chain is only as strong as its weakest link applies here. A cable system must operate in a given signal environment and distribute the signal to its subscribers. There is not much that can be done to improve the quality of the signal. Although the signal may be better than that picked up with a home antenna, it will not get any better as it passes through the system. It will, in fact, be degraded to some extent. The object of system design, operation, and maintenance is to keep the signal quality at an acceptable level. In areas where home reception is marginal at best, this is not as difficult as it is in areas where excellent off-the-air signals are available in almost every home.

The cable tv technician is usually not fortunate enough to be involved in the design of a system. More often than not, he simply inherits a system that someone else designed. Frequently, he can never be sure why the designer chose one particular approach over another. At best, his job is to keep the system operating properly, or at worst, he may have to expand its capability.

Much of the literature on cable tv abounds with descriptions of superband systems with two-way capability and sophisticated subscriber terminals. Many of the so-called MSOs (Multiple System Operators) do have these capabilities, but there are far more systems that carry only 12 or fewer channels to a limited number of subscribers. It would be impractical to add more channels or extend the length of some smaller systems. They are delivering the service for which they are designed. The technician's job is to keep them operating properly.

Many smaller systems are capable of carrying more signals or serving more subscribers. It is not always possible to predict in advance how well this can be done, but three steps in the right direction involve:

1. An up-to-date knowledge of the signal environment
2. A detailed, accurate map of the system
3. A knowledge of the present performance of the system

THE SIGNAL ENVIRONMENT

The first consideration in determining the quality of pictures delivered by a cable tv system to a subscriber is the signal environment at the headend. The signal environment is a factor that is usually not within the control of the system operator. Before a headend site is selected, a signal survey is usually made to be sure that there is enough signal strength for each station that is to be carried on the system. Equally important is the absence of interfering signals. In fact, actual site selection is usually based on the results of a signal survey.

After a headend has been installed, several things can happen that will deteriorate the signal environment. These include:

1. Changes in broadcast stations. Usually the cable system is not one of the main concerns of the broadcaster, and he may change his transmitter site or antenna system in a way that will enhance his signal in his primary service area, at the expense of the signal strength at the headend of a cable system.
2. The establishment of new co-channel or adjacent-channel stations after the cable system was installed. Under the proper propagation conditions, these new stations can cause severe interference. The same effect may result from changes in the transmitter or antenna of an existing station.
3. The erection of new structures such as water tanks in the vicinity of the headend. These structures may either reduce what used to be strong signals or cause reflections that will lead to ghosts.

In specifying the performance of a system, it is important to make sure that the signal environment is as good as it was when the system was originally installed. Many times a technician will suspect trouble in one or more components of his system when, in reality, the signal environment of the headend has changed considerably since the system was installed. The causes of some problems of this type are rather easy to isolate. When a new ghost appears in the system simultaneously with the construction of a new water tower in the vicinity, the cause of the trouble is rather obvious. The cure is often not as simple. It is sometime necessary to install a more highly directional antenna, and in extreme cases, it may be necessary to relocate the antenna.

Other problems involving the signal environment at the headend are not so easy to isolate. A system that functions beautifully in the warm months may be subject to sporadic interference from the ignition systems of snowmobiles in the winter months. At an unattended headend, troubles of this type are hard to pin down because the offending vehicle is usually gone by the time the technician gets to the site.

When co-channel interference that has previously been negligible becomes more intense, it is well to check with the operator of the co-channel station to determine if he has made a change in his transmitter or antenna that would increase the strength of his signal at the headend of the cable tv system. Sometimes problems of this type can be solved by the addition of co-channel filters.

In summary, the signal environment at the headend of the system is an important part of the overall system specifications. It is also one that can change without any of the components of the system being at fault.

THE HEADEND

The headends of two cable tv systems are never exactly the same. Some systems are fortunate enough to need only a few antennas, connected through short cable runs to a rather simple signal processor. At the other extreme, systems that are in very weak signal areas may have a preamplifier located at each of the antennas. In describing a system, we will include the components that might be found in the most extreme case. It must be remembered that in most systems many of these components are neither necessary nor desirable.

Fig. 11-1 shows a block diagram of a headend that includes most of the components that are likely to be found in a headend. The first element in any channel is the antenna. It is chosen to give enough gain for the desired signals and to give directional discrimination against undesired signals. It is much better to keep an undesired signal out of the antenna than to attempt to filter it out after it gets in.

The next block in the diagram is a bandpass filter, which is used when undesired signals are unavoidably picked up by the antenna. These signals could cause problems later on. Here again, if an undesired signal is present, it is better to get rid of it as early in the system as possible. It is important to remember that undesired signals are not only tv signals. Transmitters of two-way radio systems, and even radar systems, can put a very strong signal into your antenna if they are close by and operating at high power levels. With the rapid growth of two-way radio communication, even in rural areas, spurious signals may come from transmitters that did not even exist at the time the headend was installed. In such cases, the addition of a bandpass filter in the antenna feedline will often clean up a signal. It should be remembered, however, that even the best bandpass filter will have some insertion loss, and therefore a band-

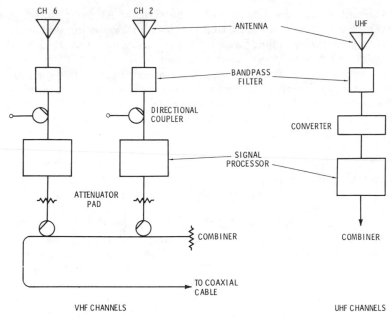

Fig. 11-1. Components of a headend.

pass filter is usually not used unless it is necessary. In addition to a bandpass filter, a band-rejection filter, or trap, is sometimes used to keep strong local signals out of other channels. In such cases, the traps are usually tuned to the sound and picture carrier frequencies of the offending signal.

The next part of the system is either a preamplifier or a converter. Let us consider the preamplifier first. As explained in Chapter 8, the signal strength is attenuated by the cable from the antenna to the signal processor, but the noise level at the input of the processor is constant. Therefore, a long cable will degrade the signal-to-noise ratio of the system. In strong-signal environments this is usually not a problem. In fact, it may be necessary to install an attenuator in the lead to keep a very strong local signal from overloading the signal processor. Nevertheless, the quality of a weak signal may be improved considerably by installing a preamplifier on the structure, close to the antenna. This increases the signal-to-noise ratio on that particular channel.

Most present-day preamplifiers are the solid-state variety, and operating power is usually transmitted to them through the signal cable. These solid-state preamplifiers are quite susceptible to overload from spurious signals and are often preceded by a bandpass filter.

199

Converters are used to handle uhf signals. As we pointed out earlier, the highest frequency that the distribution system will handle is about 300 MHz. Therefore, uhf signals must be converted to vhf or midband channels before distribution.

This frequency conversion can be handled in the headend building by the signal processor or by a converter mounted on the tower. The reason for selecting a tower-mounted converter is the same as the reason for using a tower-mounted preamplifier: it improves the signal-to-noise ratio for weak signals. Because a converter is by nature a nonlinear device, it is very susceptible to the generation of spurious signals by heterodyne action whenever two or more strong signals are present at its input. For this reason, a pole-mounted converter should always be preceded by a bandpass filter to reject undesired signals.

At the place where the signals enter the signal processor, there should be a test point at which the quality of the signals coming from the antennas can be measured and monitored, when necessary. Some signal processors have a 20-dB directional coupler built in at this point. In this case, no additional coupler is necessary. When this test point is not built into the signal processor, a 20-dB directional coupler should be installed in the line. This point is most useful for determining the quality of the signal that is entering the system, and for finding interfering signals that might cause still other spurious signals later in the system.

The signal processor itself may be any of the types described in Chapter 8. Separate agc is usually provided in the processor for the sound and picture carriers. It is customary to keep the sound carrier about 17 dB below the level of the picture carrier. This will minimize the generation of spurious signals that would result from the nonlinearities that are inevitable in any system. At the output of the signal processor, the signals are set to the frequency at which they will be distributed throughout the system. The level of spurious signals from a good signal processor may be very small, but their effects will be cumulative as they pass through the system of cascaded amplifiers. It is better to get rid of them as early in the system as possible. Therefore, it is helpful if there is a bandpass filter in each channel before the signals are combined.

EFFECT OF ATTENUATION

So far, we have thought of attenuation in a cable tv system as an inevitable evil. After all, it is because of attenuation along the cable that we need all of the cascaded amplifiers. Attenuation is not all bad, however. There are places in a cable tv system where it works to our advantage, and in some instances we actually insert attenuators in the system. The way in which attenuation is used to improve system performance can be shown by an example.

Fig. 11-2 shows the equivalent circuit of an attenuator that will introduce about 8 dB of attenuation. The impedance, Z_{in}, seen looking into the attenuator is given by the rather cumbersome expression:

$$Z_{in} = R_1 + \frac{(R_1 + 75)R_2}{R_1 + 75 + R_2} = 32.3 + \frac{(32.3 + 75)(70.9)}{32.3 + 75 + 70.9} = 75\ \Omega$$

With a little simple arithmetic, the reader can see that if the load resistance is 75 ohms, the input impedance will also be 75 ohms. With a little more arithmetic, we would find that if the load resistance changed by 10%, the input impedance would change only by slightly over 1%. In general, the greater the amount of attenuation, the less the change in input impedance for a given change in load impedance. We can, therefore, insert attenuators into a system in places where we want the impedance to appear constant. Naturally, there must be enough signal level that we can afford to do this without degrading the signal-to-noise ratio.

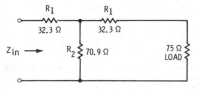

Fig. 11-2. Equivalent circuit of an 8-dB attenuator.

The fact that attenuation will make an impedance level at an input to a device remain constant even when the load impedance is varying, means that reflections from the input of the device will be minimized.

Fig. 11-3 shows how attenuation minimizes the effect of a change in load impedance on a system. The X-axis of the graph is the percentage of change in the load impedance. The Y-axis is the resulting change in the input impedance of the attenuator. Note that as the amount of attenuation increases, the change in input impedance becomes less. This principle holds not only for attenuators in which all of the attenuation is intentionally introduced at one spot in a system, but for any place in a system where attenuation occurs. Thus, the attenuation of a section of coaxial cable will tend to reduce the effect of mismatched impedances below that which would exist on a lossless line.

There are two general ways in which reflections on a cable can be held to a minimum. The first, shown in Fig. 11-4A, is to be sure that the load impedance is matched as closely as possible to the

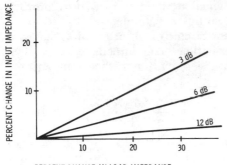

Fig. 11-3. Effect of attenuation on impedance mismatch.

characteristic impedance of the line. This will prevent reflections from occurring. In practice, loads are never perfectly matched to lines over the entire bandwidth of a cable tv system, so a second step can be taken as shown in Fig. 11-4B. Here not only is the load matched to the line, but the source impedance is also made equal to the characteristic impedance of the line. In this case, if a reflection does occur, it will be absorbed by the impedance of the source and not re-reflected back along the line. If both the source and load impedances are mismatched to the cable, the reflected signals will bounce back and forth along the cable until they are dissipated by the losses of the cable itself. This situation will provide multiple ghosts in the cable tv system, and the pictures on the subscriber's receiver will be unusable.

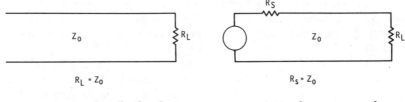

(A) *Matching load and line impedances.*   (B) *Matching source and line impedances.*

Fig. 11-4. Two ways to minimize reflections.

SIGNAL COMBINER

In the headend, it is necessary to combine the signals from each of the channels of the signal processor onto a single coaxial cable for distribution through the system. It is important that this combination be accomplished in a way that provides isolation between the various channels of the signal processor. That is, the signal from one

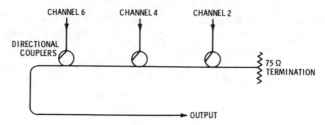

Fig. 11-5. Schematic diagram of a signal combiner.

channel of the signal processor should be kept from entering the other channels. This is accomplished by using devices that operate on the principle of the directional coupler.

Fig. 11-5 shows a diagram of a typical signal combiner. At the opposite end from the cable connectors, the device is terminated in its characteristic impedance. This ensures that any reflected signals traveling upstream will be dissipated in the termination rather than being reflected again. Signals from each of the channels of the processor are coupled to the cable through directional couplers. In the particular case shown, each directional coupler has a tap loss of 12 dB and an isolation of 30 dB. Usually, the lowest-frequency channel is connected to the distribution cable closest to the termination.

The amount of coupling between channels having the arrangement of Fig. 11-5 in signal combiners is shown schematically in Fig. 11-6. Fig. 11-6A shows the losses experienced by signals traveling in a forward direction, or downstream. Here we see that the

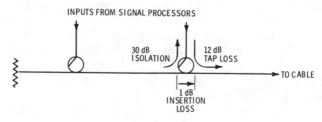

(A) Downstream losses.

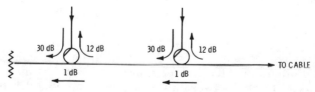

(B) Upstream losses.

Fig. 11-6. Typical losses in a signal combiner.

signal from a preceding channel will suffer a 1-dB loss as it passes through the next directional coupler. The amount of this signal that can get into the next channel from the signal processor at this point is reduced by 30 dB. The signal from each signal processor suffers a 12-dB loss upon being coupled into the cable.

Fig. 11-6B shows the losses that will be encountered by upstream, or backward, signals, which might by caused by reflections on the cable. At the directional coupler shown in the figure, a reverse signal would suffer a 12-dB loss as it passes into the signal processor and a 1-dB loss as it travels back along the cable. As pointed out earlier, each of these losses, or attenuations, will reduce the effect of the reflection.

Usually, in a practical signal combiner all the directional couplers are built into one single structure, but the principle of operation is similar to that shown in Fig. 11-5.

SYSTEM MAP

Although the headends of various cable tv systems differ considerably from one another, all headends are alike in that all of their components are located at one place. If there is a question, the technician can simply look at the headend and trace out the various signal paths. As far as the rest of the distribution system is concerned, there are even more differences between one system and another. Here, the components are not located all in one place but are spread throughout the system. For this reason, probably the most important tool that a technician can have in maintaining a cable tv system is a complete, accurate, and detailed map of the system. Usually, this map is prepared at the time the system is originally installed, but quite frequently the map is not kept up to date when changes are made to the system. Attempting to troubleshoot or maintain a system without knowing all of the details is something like trying to find a house in a subdivision with a map that does not show some of the streets.

A system map should show *everything*. The location of each pole, amplifier, and power supply should be clearly deliniated. At each distribution point along the system, a number should be placed on the map showing the number of homes that can be serviced from this point.

The chief limitation of most cable tv system maps is inaccuracies in distance. The pole-line maps used by utility and telephone companies are often quite inaccurate. A map error of 20% in a system with a 22-dB spacing between amplifiers amounts to a 4.4-dB error in signal level. This could easily place the operating point of an amplifier near one end of its dynamic range so that the signal would either drop into the noise or overload the amplifier, causing distortion.

The location of agc and mgc amplifiers should be carefully noted. The system map should show the type of service between each of the points in the system. For example, a notation might be made to the effect that there was trunk-line service only between poles 31 and 32. Of course, it is impossible to put all of this information on a single sheet of paper. The usual approach is to make an overall system map showing general information and make individual sheets showing details of the various distribution and feeder lines.

In most systems, three different types of coaxial cable are used. For reasons of economy, the cable in each part of the system is made no larger than necessary. Trunk lines, which carry the signal throughout the system, are usually made of larger cable because the larger cable has lower losses. Distribution lines, which carry signals to individual streets, housing projects, and apartments, are usually made of smaller cable. Finally, drop cables, which carry the signals into the subscribers' homes, are made of the smallest cable, usually dielectric-filled RG-type cable. Although the propagation characteristics of RG-type cable are far inferior to those of larger-type cable, the runs are usually so short that the problems involved are not too serious.

In any map of a cable tv system, the type of amplifier used at each location should be shown. Line, or trunk, amplifiers are spaced regularly along the main trunk line of the system to overcome cable loss. When a trunk amplifier happens to be at a point where distribution to other lines is required, bridging-type outputs are often included in the trunk amplifier itself. At points along the trunk line where it is not necessary to have an amplifier, but where it is necessary to branch off signals to distribution cables, bridging amplifiers are used. A bridging amplifier derives its signal from the trunk cable through a directional coupler as shown in Fig. 11-7. In the case shown, the tap loss through the directional coupler is 10 dB. This means that the signal branching off the trunk line into the bridging amplifier has $\frac{1}{10}$ the power level of the signal in the main trunk line at that point. The output of the bridging amplifier may be a single output to feed one distribution cable or may be multiple outputs to feed several distribution cables.

It must be remembered that inserting a bridging amplifier in a trunk cable actually introduces some loss in the trunk cable. Some of the signal that was in the trunk line before the bridging amplifier has been tapped off. In the example of Fig. 11-7, $\frac{1}{10}$ of the trunk-line power is fed through the directional coupler to the bridging amplifier, and 90% remains in the trunk line to continue flowing downstream. This amounts to a 0.46-dB loss introduced by the directional coupler. In addition, the directional coupler itself will have an insertion loss of as much as 1 dB. If several bridging amplifiers are added to a section of trunk line between trunk-line amplifiers, it is necessary to adjust the spacing accordingly between the trunk-line amplifiers.

The distribution cables, which are used to carry signals over shorter distances, have amplifiers also, called *line extenders*. Usually, there are fewer line-extender amplifiers cascaded in a distribution line than there are trunk amplifiers along the main trunk of the system. We have seen earlier that signal degradation, either in the form of noise or distortion, increases with the number of amplifiers that are cascaded. For this reason, the requirements for line-extender amplifiers are not as severe as for trunk-line amplifiers. Line extenders are often single-ended rather than push-pull. Ofter the line extender contains two-way or four-way signal splitters for further branching of the signal.

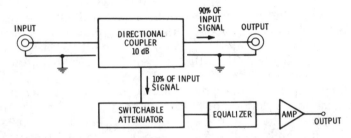

Fig. 11-7. Block diagram of a bridging amplifier.

In addition to the cable and the amplifiers, there are usually many passive devices such as signal splitters, directional couplers, and power inserters located at various points along the system. Better results are usually obtained when the passive device is contained within an amplifier housing, but for reasons of economy, this is not always possible. When the passive device is not mounted inside the amplifier housing, it should be located as close to an amplifier connector as possible, with not more than an inch or two of cable between the passive device and the nearest amplifier connector. Of course, this is not always practical either; so, when the passive device cannot be within one or two inches of an amplifier, it should be a substantial distance away from the nearest amplifier, say one or two pole spans away. The reason for these spacings is that all passive devices are mismatched to some extent over at least part of the frequency range of interest. They will therefore cause some reflection. When they are located close to an amplifier, the section of line will be so small that the reflection is not serious. When they are spaced a considerable distance from an amplifier, the attenuation of the line itself will tend to minimize the effect of the reflection.

All of these factors, including spacing, type of component, and location of each component, should be carefully marked on the system map. Above all, any change that is made on the system should be shown on the map.

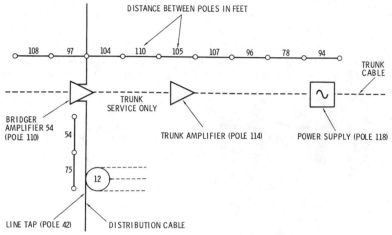

Fig. 11-8. A portion of a system map showing details.

Fig. 11-8 shows a section of a typical system map with examples of the types of notations that will be helpful both in evaluating the performance of the system and in deciding on the possible effect of proposed changes.

TAPPING

The last part of the system before the subscriber's receiver or terminal is the tap and drop cable. As pointed out earlier, any defect or limitation of a component in a cable tv system becomes more serious than it would otherwise be because of the fact that many amplifiers are cascaded in the system and each amplifier tends to increase any noise or distortion that is introduced by a component. The drop cable is the last link in the system as far as a particular subscriber is concerned. For this reason, drop cables do not have to be of the same high quality as trunk and feeder cables. This means that the drop cable is a good place to save money without seriously compromising signal quality.

Drop cables are usually the dielectric-filled RG-type. These cables are flexible and easy to route through a subscriber's home to the tv set or other terminal.

207

The drop cable may be tapped onto the feeder cable in many different ways. The best taps, and also the most expensive ones, are similar to directional couplers. Some of the earlier taps were designed more for ease of installation than for electrical performance. Many of these worked quite well on cable systems that carried only two or three channels, but when such a system was upgraded to carry more channels, it was often necessary to replace the taps.

Fig. 11-9 shows a typical modern tap. The device operates on the principle of the directional coupler or signal splitter. Twelve different tap values are provided, and these can be changed by means of an easily inserted card without removing the cable. Taps of this type can be mounted either on a pole or on a messenger cable.

Fig. 11-9. An outdoor multitap.

Courtesy GTE Sylvania Inc.

Current FCC rules provide that the minimum signal at the subscriber's terminal be 0 dBmV and that the variation between signals be not greater than 12 dB. Higher signal levels are usually available, often as high as 10 dBmV.

The FCC rules also require an isolation of at least 18 dB between the subscriber's terminal and the distribution cable. Under many conditions, such as when a tv set is sent out for repairs, the subscriber's terminals might be open-circuited. This means that 100% of the signal will be reflected back into the system. If there were no isolation, the reflections would disrupt service to all other subscribers in the vicinity. The 18-dB isolation reduces the effect of an open circuit to an acceptable level.

THE TV RECEIVER

Legally speaking, the cable tv system ends at the point where the signal enters the subscriber's tv set. The tv set itself is not part of the system. In fact, many franchise agreements expressly pro-hibit the cable operator from doing anything at all with the receiver. Nevertheless, the tv set is an essential part of the overall system. No pictures can be seen without a receiver.

Many subscriber complaints about picture quality can be traced directly to the receiver. Until recently, tv-receiver designers were concerned only with the situation in which signals were derived from a home antenna. Aside from the fact that the antenna input im-pedance is usually 300 ohms rather than 75 ohms, almost all tv receivers are designed to operate in a signal environment where co-channel and adjacent-channel interference is nonexistent, since FCC rules do not allow the use of the same or adjacent channels in the same viewing area. With cable tv, this is different. In order to obtain the maximum number of channels, the cable operator is forced to use adjacent channels. Also, direct pickup of a local station by the circuitry of the receiver can cause interference with a signal being carried on the same channel on a cable. One advantage of super-band and midband systems is that a converter is used and the subscriber's set can be left tuned to a channel where interference will not be a problem.

As cable tv grows, it is hoped that more and more manufacturers will provide tv receivers designed specifically for use on cable tv systems. Until then, the cable tv technician must find ways to live with the receivers that are available to his subscribers.

SYSTEM LEVELS

Ideally, each section of a cable tv system has a gain of unity. The signal leaves the headend at a certain level, is attenuated by the cable, and is amplified to the original level at the first amplifier. For example, in a system with 22-dB spacing, the signal might leave the headend with a level of 35 dBmV. In the first section of cable, it will suffer a loss of 22 dB, bringing the level down to 13 dBmV. At this point in the system, there will be an amplifier with a gain of 22 dB, bringing the level back to the original value of 35 dBmV.

The preceding explanation failed to mention that the attenuation of the cable is not the same at all frequencies but varies approxi-mately as the square root of the frequency. This means that the amplifier must provide more gain at channel 13 than at channel 2. Equalizers are provided in each amplifier to accomplish this task.

Obviously, if all signals are at the same level when they leave the headend, they will not be at the same level when they arrive at the input of the first amplifier. The higher-frequency signals will have suffered more attenuation and will have a lower level than the lower-frequency signals. Systems that are operated in this mode are said to have *flat* outputs. The signal levels at the output of the headend and at the outputs of each of the amplifiers are at the same level. This is not the only way that a system can be operated, and it is not the most common mode of operation. At the other extreme, the signal levels at the output of the headend and at each amplifier may be set so that the higher-frequency signals have a higher level than the lower-frequency signals. This is done so that when the signals reach the input of the next amplifier, they will all have the same level. This mode of operation is called *full tilt*. It is characterized by the fact that at the *input* of each amplifier all signals have the same level. Most modern systems operate at full tilt because it usually results in the best signal-to-noise ratio.

Still another mode of operation that is sometimes used is called *mid tilt*. This is somewhere between the two extremes of flat output and full tilt.

Although the spacing between amplifiers is based on amplifier characteristics, cable characteristics must be considered. Since larger-diameter cables have lower losses than smaller-diameter cables, a given length expressed in decibels will represent a greater physical length of the larger cable. That is, in a low-loss cable it will take a greater number of feet of cable to produce the same loss in decibels as a shorter length of cable having a higher loss. For this reason, main trunk cables usually use large-diameter cables that have low losses. In these trunk cables there will be a large number of cascaded amplifiers, and use of the lower-loss cable will permit a longer system for the same number of amplifiers. Feeder cables, which will not be as long, can use smaller cable with higher losses. The actual selection of the type of cable to be used in each part of a system is a compromise between the cost of the cable and the cost of the amplifiers.

NOISE AND INTERFERENCE

One of the problems that occurs in any mode of tv reception, either cable or off-the-air, is noise. Every electrical device will generate noise. Theoretically, the least amount of noise that can exist in any system is thermal noise, which was discussed in an earlier chapter. In practical devices, the noise will be much greater than this. It is a fact of life that there will be noise in a cable tv system and that the input stage of the subscriber's receiver will introduce still more noise. The question that arises is just how much noise can be present in a system before the picture is seriously degraded.

It must be remembered that picture quality is very subjective. Whether or not a given picture is acceptable depends on whether or not the viewer finds it to be acceptable. Because of this, many different groups have conducted tests with representative viewers to find out just how much noise, interference, and distortion they will tolerate in a tv picture. One of the most extensive studies of this type was made by the Television Allocations Study Organization (TASO) in 1959. The results of this study are still used as guidelines by the designers of cable tv systems.

Fig. 11-10 shows a plot of the TASO findings regarding the amount of noise that viewers found acceptable. The curves show the percentage of viewers that found certain signal-to-noise ratios to be inferior, marginal, passable, fine, or excellent. Obviously, in a metropolitan area where many excellent signals can be picked up on a home antenna, a system must have excellent pictures in order to be competitive. In a rural area where the pictures obtained with a home antenna are marginal at best, the cable system can get away with pictures that would not be acceptable in the metropolitan area.

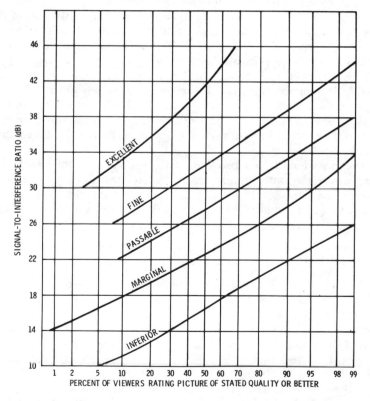

Fig. 11-10. Effect of various levels of noise as judged by 38 male and 38 female observers.

To some extent, the amount of noise that will be tolerated depends on the nature of the noise itself. In general, low-frequency impulsive noise is more objectionable than higher-frequency noise that will produce fine snow in the picture. One limiting factor in the amount of noise that can be tolerated in a cable tv signal is the noise figure of the subscriber's tv receiver. The signal-to-noise ratio is effectively reduced by the noise figure of the receiver. The 1959 TASO study showed that the average noise figures for home receivers available at that time were 5.5 dB for channels 2 through 6 and 7.5 dB for channels 7 through 13. Many receivers in use today have noise figures that are not much better than this. As receiver noise figures are reduced, better pictures will be produced from both home antennas and cable systems.

So far, we have talked only about noise that is pickeed up by the headend or generated in the system. Although the coaxial cable and amplifier housings are supposed to be "airtight" as far as outside signals are concerned, in practice the shielding is often far from perfect. This gives rise to two problems: First, the cable system may pick up signals from the outside; these will cause interference and degrade picture quality. Second, the cable system may radiate signals that interfere with other services.

These problems of pickup and radiation are most apt to cause trouble in systems that use the midband channels. Many two-way radio services operate in this part of the spectrum. Interfering signals from mobile units are particularly hard to track down because the mobile unit will often have left the scene before the investigator arrives.

Pickup and radiation problems can usually be traced to lack of integrity of the outer conductor of the system. Such problems are often caused by amplifier housings or by connectors that are old and have been attacked by weather. Techniques for locating radiation and pickup are treated in a later chapter.

PROVIDING MORE THAN TWELVE CHANNELS

Most of the earlier cable tv systems were designed with a single trunk and amplifiers that carried channels 2 through 13. Where no strong local signals were present and where the second-order distortion of the system was low, as many as 12 channels could be provided by such a system. These systems were entirely acceptable in localities that had essentially no home reception without the cable. In many areas they are still acceptable.

In the major markets it is a legal requirement that cable systems carry more than 12 channels. In many of the smaller markets there is subscriber pressure for more channels. So, new systems are being designed to carry at least 20 channels and in many instances even more. When properly designed and adjusted, they present no major problems. Updating older systems to carry more channels is, unfortunately, a different matter. There are three ways to get a channel capability greater than 12:

1. The midband frequencies between channels 6 and 7 may be used. This requires a converter at the subscriber's terminal because most present-day receivers cannot be tuned to the midband channels.
2. The super-band channels above channel 13 may be carried. This also requires a converter.
3. A second cable may be run along the entire system to provide up to 12 additional channels. This is the *dual cable* or *A-B* system.

There is no best way to update the channel capability of all systems. Much depends on the quality of the original system and the density of subscribers in each portion of the system. Adding midband signals should not even be considered unless the second-order distortion of the original system is low. Remember that the effect of second-order distortion is to produce new signals at frequencies equal to the sums and differences of the signals carried on the system. When only a few signals are carried on channels 2 through 13, a great deal of second-order distortion can be tolerated in a system. However, when most of channels 2 through 13 are used and the midband channels are also carried, the system cannot tolerate any appreciable second-order distortion. The system will, in effect, create its own interference. The best approach to minimizing second-order distortion is to use push-pull amplifiers throughout the system.

If the second-order distortion of a system is low to start with, adding the midband channels is probably the best way to expand the capability of the system so that it can carry more channels.

Updating a system by replacing all of the trunk and distribution amplifiers should be approached with caution. In some older systems that use vacuum-tube amplifiers, the spacing is so large that a considerable amount of cable splicing will be required in order to use new wideband amplifiers. This will usually result in a system that is very hard to clean up.

One problem that sometimes occurs with midband systems is that the signal from the local oscillator of one subscriber's converter may travel along the cable to another subscriber's converter. This is sometimes referred to as two converters "talking to each other." The converter-to-converter interference problem is shown in Fig. 11-11. The converter is similar to the converter in a tv set. It converts the

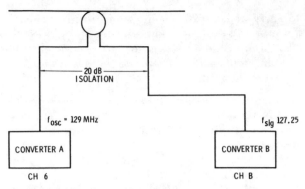

Fig. 11-11. Converter-to-converter interference.

signal from the input to an i-f of 45.75 MHz and then converts the i-f to the desired channel. In the figure, converter A is tuned to channel 6. Its local oscillator thus operates at 129 MHz. If converter B is tuned to this frequency in the midband, the signal from the local oscillator of converter A will deliver an interfering signal. In this example, there is 20 dB of isolation between each converter and the cable; however, if the level of the local oscillator signal is 0.5 V, which is a reasonable level, the interfering signal will have a level of +34 dBmV.

The problem of converter-to-converter interference can be eliminated by using a converter that has a very high i-f, say above 300 MHz. The local-oscillator signal will then be above any of the tv channels.

This type of problem is most severe in so-called mixed systems. A mixed system is one to which the midband or super-band channels have been added long after the system was originally built. In such cases it is customary to give the subscriber the option of whether or not he wants to install a converter, usually at a higher monthly rate. As a result of this arrangement it is possible, in fact even probable, for one subscriber to have a converter while his neighbor does not have one. In such a case the local oscillator of the tv set without the converter may easily radiate a signal into one of the midband channels.

In a system where the subscriber is happy with reception from the cable only, a low-pass filter can be installed at his receiver terminals. However, many subscribers insist on the option of using either the cable or their own antenna. In such cases, the only solution is to introduce additional attenuation in the form of one or more directional couplers.

The dual-cable system, as its name implies, has two separate coaxial cables, each with its own amplifiers. This approach is used for updating existing systems and is also used in many new systems to provide a capacity for carrying many channels. This method has much to recommend it for updating existing systems. First of all, the original system (which, we hope, is operating properly) can be left intact. Second, if the original system has many limitations, the second cable can have greater capabilities. Then later, perhaps, the original cable can be brought up to the quality of the second cable. The dual-cable approach to updating a system allows uninterrupted service to subscribers and need not be carried out all at one time.

The dual-cable system is very similar to a single-cable system except that all of the amplifiers and other components distributed along the system are duplicated. In some instances when a new system is installed with two cables, an attempt is made to use common power supplies for the amplifiers in both cables. This can be done successfully, but a great deal of care must be exercised to prevent cross coupling of signals from one cable to the other.

CHAPTER 12

Two-Way
Transmission and
Special Services

Cable tv was developed to bring tv programs to areas where off-the-air reception was impossible. Originally, systems were designed only to relay regular tv broadcast signals. In a broader sense, a cable tv system is a broad-band communications network. Operators have found that they can increase the number of subscribers by adding special services far beyond the relaying of tv broadcasts.

The basic cable system is designed to transmit large blocks of information—tv programs—from a central location to many locations. As far as the cable itself is concerned, the information could flow either way. With modifications, amplifiers can also carry signals either way. The two-way capability makes possible almost unlimited applications for the cable system.

One obvious application is home merchandising. An advertiser can display his products by cable, and the subscriber can order by merely entering his credit-card number on a home terminal. Sizes, quantities, and other details of the purchase could be entered into the system by means of a keyboard. An offshoot of this is home billing; a subscriber could have any of his bills sent to him directly on his tv screen. A further application of this system would be automatic reading of utility meters. Once a two-way capability is provided, the number of possible applications seems to be limited only by the imagination.

The main problem in providing a two-way capability in a cable tv system is that the amplifiers in the system are one-way devices. They are designed to handle the regular downstream tv signals and have no way of handling upstream signals traveling from the subscriber's terminal to the headend.

The usual way of providing upstream transmission is to use frequencies below those normally carried by the system for the upstream signals. At each amplifier location, an arrangement such as that shown in Fig. 12-1 is used to separate the upstream and downstream signals. High-pass filters keep the downstream signals in the regular amplifiers, and low-pass filters keep the upstream signals in separate amplifiers.

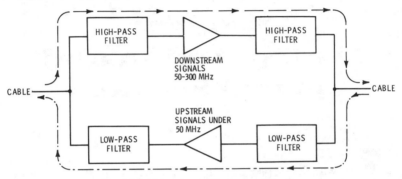

Fig. 12-1. Two-way cable-tv amplification.

The arrangement of Fig. 12-1 appears very simple, but the simplicity is deceptive. The filters can cause many problems if they are not designed and adjusted properly. In fact, keeping a two-way system working properly involves a good feeling for the ways in which the filters can distort the signals. Before going further, we will review some of the basic principles of filters.

A FEW FACTS ABOUT FILTERS

Before discussing the properties of filters that influence the behavior of a two-way cable tv system, we will review how filters work. Fig. 12-2 shows a low-pass filter. It is designed to pass all signals at frequencies below 50 MHz with little attenuation and to reject all frequencies above 50 MHz. Although we are looking at this filter as a one-way device with the signal entering at the left and leaving at the right, we can actually look at it as a two-way device as long as it has only resistance, inductance, and capacitance and contains no active components such as transistors or integrated circuits. That is, the filter will also pass frequencies below 50 MHz

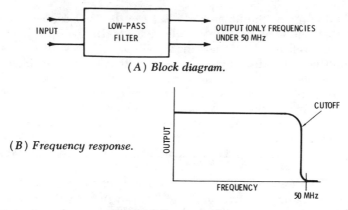

(A) Block diagram.

(B) Frequency response.

Fig. 12-2. Low-pass filter.

and reject frequencies above 50 MHz if the signals are applied at the right and leave at the left.

We can get a better idea of how the filter works by looking at its input impedance at various frequencies. Fig. 12-3 shows the filter connected to a 75-ohm load. As shown in the figure, at frequencies below 50 MHz the input of the filter looks like a resistive impedance of 75 ohms. Another way of saying the same thing is that as far as frequencies below 50 MHz are concerned, the signal does not see the filter at all. All it sees is the resistive load connected to the output terminals of the filter. This is why the word "transparent" is sometimes used in connection with filters.

We have stated above that our filter will reject all frequencies above 50 MHz, but we have not defined exactly what we mean by the word "reject." What happens at these frequencies is that the input impedance of the filter is no longer the resistive impedance of the load, but is nearly a pure reactance. We know that no real power enters a reactive load, so the question is, "What happens to a signal having a frequency above 50 MHz when it reaches the input of the filter?" The answer is that this signal is reflected back toward the source. The filter is not transparent at these frequencies.

We might describe the behavior of our filter by saying that at frequencies below 50 MHz it looks like a transparent piece of glass.

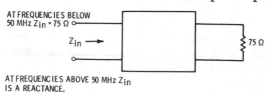

Fig. 12-3. Input impedance of a low-pass filter.

Signals do not see the filter at all, but see the load connected to its output terminals. At frequencies above 50 MHz, the filter looks like a mirror. It reflects the signals that reach it back toward the source.

From the above discussion, we can see that a low-pass filter does to signals above its cutoff frequency one of the things we dread most in a cable tv system: it causes reflections. At first glance, it would appear that this would rule out filters in cable tv systems, but there are design techniques that will alleviate the problem. The important thing to bear in mind is that filters can and do cause reflections, and this must be taken into consideration in cable tv systems.

Fig. 12-4 shows a section of a two-way cable tv system at the input of a regular downstream amplifier. At present, we will ignore the upstream signal and see what the filter arrangement does to the regular downstream signals. For the moment, the high-pass filter is no problem. (There are other considerations that we will discuss later.) In parallel with the high-pass filter is a low-pass filter. Its purpose is to pass upstream low-frequency signals onto the cable. The regular downstream signals will, however, be reflected from its terminals.

Earlier we mentioned that as far as the high-frequency downstream signals are concerned the low-pass filter looks like a reactance. The trick is to make this reactance high enough at the frequency of the downstream signals that the combined impedances of the two filters will look very much like a pure resistance. As shown in Fig. 12-4, the combined parallel impedance of a 75-ohm resistive load and a 1000-ohm reactance amounts to 74.5 + j5.6 ohms.

In the above discussion, we have assumed that we could get low-pass and high-pass filters that had an abrupt change in their amplitude response at the cutoff frequency. In practice, the transition from the pass characteristic to the reject characteristic is more gradual, as shown in Fig. 12-5. By adding more sections to the filter, we can make the characteristic steeper, but we cannot do this without paying a price.

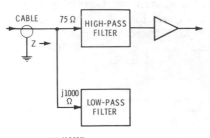

Fig. 12-4. Input section of two-way amplifier.

$$Z = \frac{(75)\,(j\,1000)}{75 + j\,1000} = 74.5 + j\,5.6 = 74.8 \angle 4.2^{0}$$

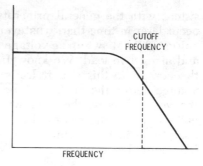

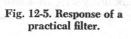

Fig. 12-5. Response of a
practical filter.

It is a fundamental principle of circuit theory that you cannot change the amplitude response of any circuit element, including a filter, without also changing the phase response. The sharper the change in the amplitude response as a function of frequency, the more drastic the change in phase shift. (You can build circuits that will change the phase shift without changing the amplitude response, and this is often done to compensate for the phase properties of a filter.)

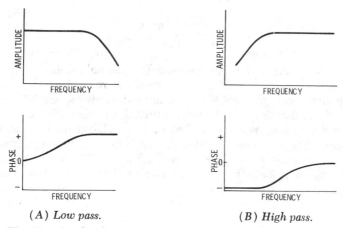

(A) Low pass.   (B) High pass.

Fig. 12-6. Amplitude and phase characteristics of practical filters.

Fig. 12-6 shows the amplitude and phase response curves of both a high-pass filter and a low-pass filter as a function of frequency. Note that the output of a low-pass filter lags the phase of the input, whereas the output of a high-pass filter leads the phase of the input. The subject of a leading phase angle is often confusing. It means that the output voltage reaches the maximum part of the cycle before the input voltage. This seems as though the effect is anticipating the cause, which, of course, it cannot. Before going further, we will clarify just what a leading phase angle is and how it is consistent with the general principle that any effect must necessarily occur later in time than whatever causes it.

Fig. 12-7A shows an ac voltage source, a switch, a series capacitor, and a resistive load. We know from elementary circuit theory that the current in this circuit leads the applied voltage and that the voltage across the load must be in phase with the current. Thus, the voltage across the load will reach its peak value before the voltage across the source. Inasmuch as the source is what causes everything in the circuit, it superficially appears that the effect is happening before the cause.

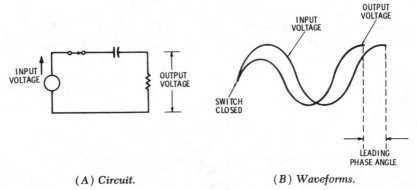

(A) Circuit.   (B) Waveforms.

Fig. 12-7. Leading phase angle in an RC circuit.

We need not go into a mathematical analysis; we will ignore some of the details of the transient period, but they are not essential to our discussion. We must start with no charge in the capacitor. Suppose that we close the switch just when the voltage from the source is passing through zero, as in Fig. 12-7B, A capacitor with no charge in it looks like a short circuit the instant that a voltage is applied to it; so, as the voltage from the source starts to rise, all of the voltage will appear across the resistive load. So far, the cause and effect are happening at about the same time. But as the voltage from the source starts to rise, the capacitor will start to charge. This means that the voltage across the load resistor will not rise as fast as the voltage from the source. Now the effect is happening after the cause, which is what we should expect.

What is happening in the circuit is that the capacitor is trying to charge to the voltage of the source. A point will be reached where the voltage across the capacitor is still rising, but the voltage across the load is actually decreasing. Once the transient period is over, the voltage across the load will in fact reach its peak value before the voltage from the source does. This is not a case of the effect anticipating the cause, but is a result of the transient period.

222

If we were to mark one of the cycles of the source voltage with a little glitch, as in Fig. 12-8, the glitch would appear at the same time in the output. Thus, a leading phase angle is not the same thing as an advance in time. Signals will never leave any circuit element, unless it is oscillating, before they enter it.

In any case, for a tv signal to be undistorted, the phase versus frequency response of any filter, high-pass or low-pass, must be a linear function of frequency.

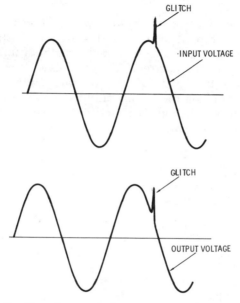

Fig. 12-8. Leading phase angle does not mean negative time delay.

PRACTICAL TWO-WAY AMPLIFIERS

Fig. 12-9 shows the block diagram of a practical two-way amplifier. It consists essentially of two amplifiers. The downstream amplifier handles the regular tv channels between 50 and 300 MHz. The upstream amplifier handles the upstream signals which occupy that part of the spectrum between 5 or 6 MHz and about 25 or 26 MHz. There are no signals between 25 and 50 MHz. The amplitude and phase characteristics of the filters are worst in this frequency range, but since the system handles no signals here, there are no problems. The signals are kept in their proper paths by high-pass and low-pass filters which are usually called *diplexer filters*.

Tests have shown that if the color and luminance information of a tv picture are out of step by as much as 50 nanoseconds, the tv set cannot resolve the difference. The time period corresponding to a degree of phase error is about 0.9 nanosecond per degree. Thus, a total phase error in a system of as much as $50/0.9 = 55°$ is usually not noticeable. If 50 amplifiers are cascaded, however, the maximum permissible phase error on this basis will be less than 1.2 degrees for each amplifier. This shows how critical the phase properties of the filters are.

UPSTREAM SIGNALS

No discussion of a two-way system would be complete without a mention of the nature of the signals that are sent upstream from the system to the headend. The possibilities are almost unlimited. So far, the signals used in the upstream direction have been of two types. One is simply a tv signal. Local programs may be originated at almost any point along the trunk, and in some cases at any point in the distribution system, and sent back to the headend for retransmission on the proper channel.

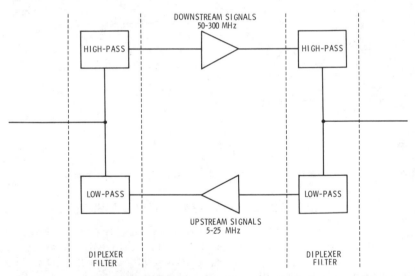

Fig. 12-9. Practical two-way amplifier.

The other class of signal that is often used in the upstream channel is a digital signal that will both carry information and identify the source of the signal. An example is an intrusion alarm service. If an intrusion occurred at a subscriber's home, the digital signal would be sent back upstream, indicating that an intrusion had occurred and identifying the home. Such digital signals could also be used to read utility meters and to indicate which channels were being viewed.

Still another use of the upstream channels which appears very promising is the monitoring of the parameters of the cable tv system itself at various remote locations in distant parts of the system. A system with this feature would provide an automatic indication of trouble and could be used to help pinpoint the location of the faulty part of the system. It would also provide the background informaion needed for a fully automatic, computer-controlled cable tv system.

PAY TV OR EXCLUSIVE PROGRAMMING

Two types of programs that are of increasing interest are pay-tv programs (where a subscriber could see a special program, such as a first run movie, for an extra fee) and programs that are directed toward a limited group of viewers. The latter category could include materials such as surgical techniques that were intended for surgeons but might not be considered suitable for viewing by the general public.

Exclusive or restricted programming is handled by scrambling the signal so that neither the sound nor the picture can be recovered unless the subscriber is equipped with a descrambler. There are several ways that the sound of a tv signal can be altered so that it is not intelligible on an ordinary tv set. One way is to scramble the sound channel just as speech is scrambled in secure communications systems. Another way is to shift the carrier frequency of the sound so that it will not be picked up by a normal tv receiver.

The video information in a picture signal can, of course, be scrambled also, but since the timing of a tv signal is critical, the design and adjustment of such a scrambler and the corresponding descrambler are critical. Another method that is used to make signals unviewable on an ordinary tv set is the removal of all or part of the sync pulses. Fig. 12-10A shows one line of a video signal. Fig. 12-10B shows the same signal clamped to a gray level so that there are no sync pulses for the receiver to lock onto. Such a signal would not produce an intelligible picture on an ordinary tv set. The sync pulses that are missing from Fig. 12-10B are transmitted on another carrier within the 6-MHz passband. A converter placed near the set detects the sync pulses and restores them to the signal. Thus, the picture can be viewed only by subscribers who have the proper descrambler.

The scrambler can be connected to an upstream channel to provide an indication at the headend of which subscribers watch a premium program so that they can be charged for the service.

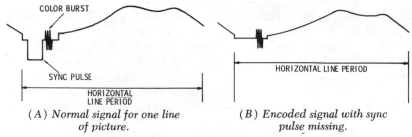

(A) *Normal signal for one line of picture.*

(B) *Encoded signal with sync pulse missing.*

Fig. 12-10. Encoding of tv signal.

FRAME GRABBERS

Many of the special services that are well suited for a cable tv system require that material be displayed on the screen of a tv set long enough for a subscriber to read the material. For example, one such service might be a page or two of stock market quotations. The subscriber could dial for the service by using the upstream channel of a two-way system, and the information could be sent on one of the channels carried by the system.

The regular method of transmission would be extremely inefficient for this or similar applications. What is needed is a way to send the information for a short period of time and somehow store it at the subscriber's terminal. Devices that have been designed for this purpose are often called *frame grabbers*. The picture desired by the subscriber is sent for 1/30 or 1/60 of a second and is stored at the subscriber's location as long as he wishes.

Frame grabbers have been built that use storage tubes to store the field or frame by means of electrostatic storage techniques. The rapid advance of semiconductor memories and digital devices such as time-base correctors is making it possible to use a semiconductor memory for the purpose.

One channel carried by a system could be devoted to such special services. Each page could be identified by a special signal in the vertical-blanking interval. If the subscriber set his controls for a particular series of frames, he would receive them at a rate of one every ten or twenty seconds, with the memory holding each frame until the next frame of interest was transmitted. In this way, the special channel could transmit hundreds of different programs, with the subscriber selecting the ones in which he was interested.

CHAPTER 13

Long-Distance Transmission

In all of our discussions thus far, the cable system has been intended to transmit tv signals from the headend to the subscriber's receivers. It has been tacitly assumed that all of the signals were available at the headend, and that the distance from the headend to the most distant subscriber was not so long as to cause great difficulty. Unfortunately, this is not always true. There are situations where it is necessary for the cable tv system to transmit tv signals over distances that are so great as to make the conventional cable transmission system impractical. In these situations, the signals are transmitted either by a microwave transmission system or by a somewhat different type of cable transmission system which is usually called "supertrunk."

One application of long-distance transmission is illustrated in Fig. 13-1. The headend is located so far from the area to be serviced that the attenuation of a regular cable system would be excessive. In the arrangement shown, the signals are relayed from the headend to a central distribution point, called the *hub* of the system. From the hub, the signals are distributed to the subscriber's homes just as they would be in any other cable system.

In another application of long-distance transmission, a single headend serves several communities, as shown in Fig. 13-2. Here the off-the-air signals are picked up at the headend and are transmitted by an omnidirectional antenna so that they can be picked up by hubs in other communities. Still another use of long-distance

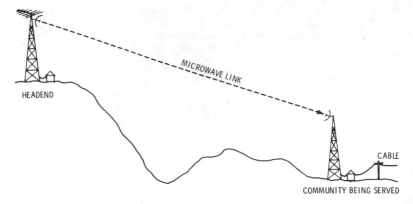

Fig. 13-1. Microwave link from remote headend.

transmission is in picking up one or more distant signals at some location other than the headend of the system and transmitting the signal to the headend by a microwave link.

In all cases, some mode of transmission other than that normally used on the cable is used to minimize the losses over the long path. This is accomplished by using some frequency other than the normal tv channel frequencies for transmission. These special modes of transmission, although they have lower losses, are not practical for

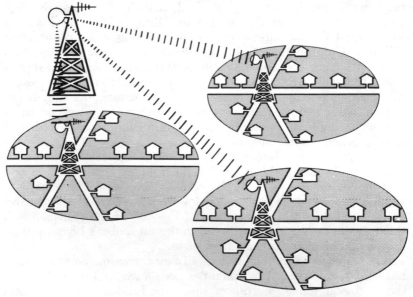

Fig. 13-2. Microwave interconnection of cable tv systems.

the main cable tv system because they require expensive frequency-conversion equipment at the subscriber's home.

MICROWAVE TRANSMISSION

All microwave transmission systems must be licensed by the FCC. In the FCC nomenclature, microwave systems used in connection with cable tv systems are classified as being in the Cable Television Relay Service. Frequencies in the 12,000-MHz portion of the spectrum have been set aside for this service.

The types of stations recognized by the FCC include the following:

1. *Cable Television Relay Station.* This type of station is often referred to as a CAR station. This is because the service was formerly called the Community Antenna Relay Service (CARS). A cable tv relay station is any fixed or mobile station that picks up signals and then transmits them to a terminal point from which they are distributed to the public by cable.

2. *Local Distribution Service (LDS) Station.* This is a CAR station located within a cable tv system and used for transmission to one or more locations from which the signals are distributed to the public by cable.

3. *Cable Television Relay Studio to Headend Link (SHL) Station.* This is a fixed station that transmits signals from a studio to the headend of a system. It is comparable to the studio-to-transmitter link (stl) used in radio and television broadcasting.

4. *Cable Television Relay Pickup Station.* This is a mobile station that is used to pick up programs from a location other than the studio. The signals are transmitted to the headend of the system.

A microwave system uses a carrier in the 12,000-MHz portion of the spectrum to carry the tv signals. Either amplitude or frequency modulation may be used. Usually, parabolic antennas having high gains are used to enhance the signal strength.

Propagation and signal behavior at microwave frequencies are not quite the same as at tv frequencies. There are just enough differences to cause trouble for the engineer who is not familiar with them. We will, therefore, review some of the principles of microwave propagation.

PROPAGATION OF MICROWAVES

As the frequency of a radio signal becomes higher, the behavior of the waves more closely resembles that of a light beam. Signals

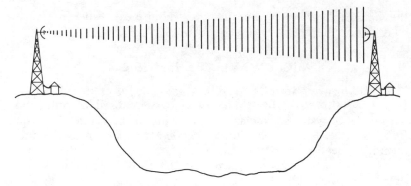

Fig. 13-3. Unobstructed path for microwaves.

travel in direct, line-of-sight paths and can, for many purposes, be treated like rays. To some extent, this is true of tv signals, but the correspondence is much closer with microwaves. Like tv signals, microwaves can be reflected, refracted, or diffracted. Diffraction tends to be the most mystifying phenomenon until it is properly understood.

Fig. 13-3 shows a microwave link with a clear, unobstructed path between the transmitter and receiver. Fig. 13-4 shows the same transmitter and receiver, but in this instance there is an obstruction such as a building or hill in the path. Although the obstruction does not block the line-of-sight path between the transmitter and receiver, the signal strength is reduced by 6 dB. We just said that microwaves travel in a direct line-of-sight path, and now we are saying that when some of the rays that were not aimed at the receiving antenna are blocked, the received signal will be reduced. This deserves further explanation.

The phenomenon known as diffraction is best explained in terms of a principle from physics known as Huygens' principle. The principle was originally applied only to light, but it applies

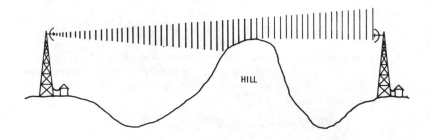

Fig. 13-4. Transmission path partially obstructed.

equally well to microwaves which behave much like light. Huygens proposed that propagation of light could be explained if every point on an advancing wavefront were considered to be a secondary source of radiation. Fig. 13-5 illustrates the principle. Here a mask is placed between a light source and a screen. If a pinhole is made in the mask, the whole screen will be illuminated by the light passing through the pinhole. The strongest light will be at point P, which is directly behind the pinhole.

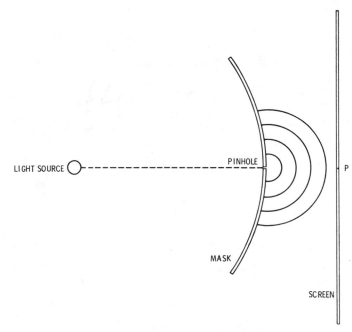

Fig. 13-5. Illustration of diffraction.

In Fig. 13-6, we have the same situation, but now there are two pinholes in the mask. Here it can be seen that the illumination at point P is received from both pinholes. Since the light travels by two different paths between the source and point P, the two light beams will tend to either cancel each other or reinforce each other, depending on the relative lengths of the two paths. If the difference in path lengths is an odd multiple of a half wavelength, the two signals will be out of phase and will tend to cancel. If the difference in path lengths is an even multiple of a half wavelength, the signals will be in phase and will tend to reinforce each other.

Now, suppose that instead of the two pinholes shown in Fig. 13-6, we had a very large number of pinholes. The illumination of the screen at point P would contain a component from each pinhole. From a large number of pinholes, it is an easy step to add an infinite number of pinholes. This is the same as removing the mask completely. This shows that although the light travels in a straight line, the illumination at point P depends on light from the entire wavefront. Most of the illumination comes from the direct line from the light source, but some of it comes from each part of the wavefront.

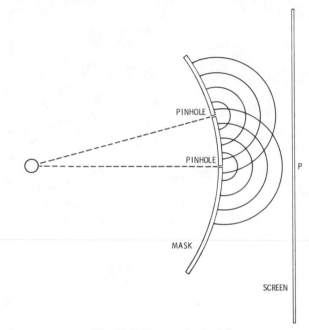

Fig. 13-6. Huygens' principle.

We stated earlier that Huygens' principle applies to microwaves as well as to visible light. Thus, when there is an obstruction in the path of a microwave signal, the signal strength will be reduced even though the obstruction does not block the line-of-sight path. By the same reasoning, some signal may be picked up behind the obstruction when there is no line-of-sight path, although the signal strength will be much lower than it would be in the direct path.

The above discussion naturally leads to the question of how much of the wavefront can be blocked without seriously interfering with the signal strength. A quantitative answer to this question involves mathematics beyond the scope of this book. A qualitative explanation is shown in Fig. 13-7. Here the wavefront is shown broken up into zones which are called *Fresnel* (pronounced Fray-

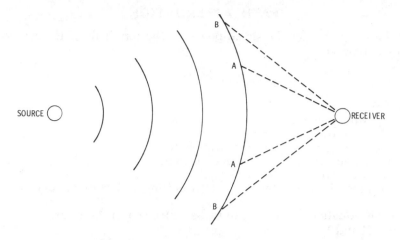

Fig. 13-7. First and second Fresnel zones.

nell') *zones* after their discoverer. The first Fresnel zone extends from the center of the wavefront to points A. The signals across this portion of the wavefront are in phase and will tend to add. The second Fresnel zone lies between points A and points B. Signals from this portion of the wavefront tend to cancel those from the first zone because of the difference in path length.

Most designers of microwave links try to get first Fresnel zone clearance, although clearance of one half of the first Fresnel zone is adequate.

FADING

In all but very short microwave links, there is some fading of the signal. The principal cause of fading is the fact that at microwave frequencies, there is a great deal of absorption and scattering from moisture and precipitation. Where there are obstructions in the transmission path, refraction due to atmospheric conditions may bend the signal so much that a large portion of the wavefront is obstructed.

The amount of fading that will occur over any particular path can be determined only by actual experience. However, by allowing a safety margin, very reliable service can be provided. Designers usually specify the fading along a path in statistical terms. If a system has a fading probability of 0.1%, the signals will be unusable only about 9 hours per year. Reliable systems are completely out of service for a total of less than one hour during a year as a result of fading.

233

PATH ATTENUATION

The equation for the signal power at the terminals of the receiving antenna is:

$$P_R = \frac{P_T G_T G_R \lambda^2}{(4\pi R)^2}$$

where,

P_R is the power at the receiving antenna,
P_T is the transmitter power,
G_T is the gain of the transmitting antenna,
G_R is the gain of the receiving antenna,
λ is the wavelength of the signal,
R is the distance between the transmitting and receiving antennas.

In this equation, P_T and P_R must be expressed in the same units, as must R and λ.

In order to keep this equation compatible with those we used earlier to express the transmission loss in coaxial cable, we will convert it to logarithmic units. The equation then becomes:

$$P_r = P_t + G_t + G_r - 20 \log 4\pi R + 20 \log \lambda$$

where,

P_t is the transmitter power in dBm (that is, in decibels with respect to one milliwatt),
P_r is the power level in dBm at the receiving antenna,
G_t is the gain of the transmitting antenna in dB,
G_r is the gain of the receiving antenna in dB.

The other symbols have the same meaning as in the preceding equation. R and λ must be expressed in the same units.

The last term of this equation is expressed in wavelengths, which are not as familiar as frequency. We can convert from wavelength to frequency by using the relationshsip:

$$\lambda = \frac{300}{f_{MHz}}$$

The equation now becomes:

$$P_r = P_t + G_t + G_r - 20 \log 4\pi R - 20 \log \frac{f}{300}$$

where f is the frequency in megahertz.

The last term of this equation would seem to indicate that losses were higher in systems using higher frequencies. This is misleading, however, because if the receiving antenna has a given area, its gain will increase with frequency so that for all practical purposes the losses are independent of frequency.

The next-to-last term of the equation gives the loss due to the length of the path. From this term, we can see that the loss will increase by 6 dB every time the distance between the transmitting and receiving antennas is doubled. This is in sharp contrast to the loss in a coaxial cable, which is a constant number of decibels per unit of length. Thus, if the length of a coaxial cable is doubled, the loss in decibels will be doubled, whereas in a microwave link the loss will be increased by only 6 dB.

In microwave systems, power levels are usually expressed in dBm, that is, in decibels using one milliwatt as the zero reference level. In cable tv systems, it is customary to express power levels in dBmV, that is, in decibels with one millivolt across 75 ohms as the zero reference. To convert from one expression to the other:

1. To get dBm, subtract 48.75 dB from the number of dBmV.
2. To get dBmV, add 48.75 dB to the number of dBm.

Before we go on, let us put some numbers into our equation for signal strength in a microwave link. Assume that the transmitter power is one watt ($+30$ dBm), the gains of the transmitting and receiving antennas are both 40 dB, the frequency is 12,825 MHz, and the path length is 30,000 meters (about 19 miles). The equation becomes:

$$P_r = + 30 + 40 + 40 - 20 \log 4\pi \times 30{,}000 - 20 \log \frac{12{,}825}{300}$$
$$= + 30 + 40 + 40 - 111.5 - 32.5$$
$$= - 34 \text{ dBm or } 14.75 \text{ dBmV}$$

RECEIVER THRESHOLD

The minimum signal strength that can be allowed at the receiving end of a microwave link depends on the noise level at the input of the receiver. This noise power level in dBm is given by:

$$N = 10 \log KTBN_f + 30$$

where,
N is the noise power level in dBm,
K is Boltzmann's constant, 1.37×10^{-23} joule per kelvin,
T is the temperature in kelvins (293 kelvins at room temperature),
B is the bandwidth of the microwave receiver in hertz,
N_f is the noise factor of the receiver.

In a practical case, where the bandwidth is about 15 MHz and the noise factor of the receiver is 12.6 (11 dB), the noise power in dBm works out to be -92.2 dBm.

If this type of receiver were used with the link described in the preceding paragraphs, the receiver signal-to-noise ratio would be:

$$S/N = -34 - (-92.2) = 58.2 \text{ dB}$$

This ratio is frequently called the a-m carrier-to-noise ratio. The reason for specifying the signal-to-noise ratio in terms of an a-m signal will become apparent when we discuss frequency modulation of the carrier.

There is often a great deal of confusion regarding signal-to-noise ratio in a complete receiving system. In addition to the bandwidth of the rf and i-f sections, the bandwidth of the video amplifier following the detector must be taken into consideration to determine the overall signal-to-noise ratio.

The subject of signal-to-noise ratio in microwave links is confusing because some types of modulation have a signal-to-noise advantage. Under some conditions, an fm signal is not as susceptible to noise as an a-m signal. In evaluating the noise performance of a given signal, the bandwidth of the video amplifier following the detector in the receiver must be taken into consideration.

A-M RECEIVER PERFORMANCE

In the preceding paragraphs, we computed the noise level in a receiver. The signal-to-noise ratio that we calculated was the ratio of the carrier to the root-mean-square (rms) noise level in the receiver. What is more pertinent is the ratio of the video signal to the noise level at the output of the receiver. In order to compute this figure, we must make a few additions to our equations. It is common to express the video signal-to-noise ratio in terms of the peak-to-peak video signal and the rms value of the noise. The peak-to-peak value of a video signal is about 9 dB greater than the rms value. This means that we must add 9 dB to the carrier-to-noise ratio that we found earlier.

In addition, there is another factor that will improve the video signal-to-noise ratio. This is due to the fact that the bandwidth of the video stages of the receiver is much narrower than that of the rf stages. Actually, both bandwidths enter into the equation. The amount that the signal-to-noise ratio, expressed in decibels, must be increased due to bandwidth considerations is:

$$10 \log \frac{B_{rf}}{2B_v}$$

where,

B_{rf} is the bandwidth of the rf and i-f portions of the receiver,
B_v is the bandwidth of the video portion after the detector.

The video bandwidth is doubled because in the detection process, noise is taken from the portion of the spectrum containing the signal and also from what amounts to an image frequency.

The equation for the ratio of the peak-to-peak video signal to the rms noise then becomes:

$$\frac{S}{N} = \frac{C}{N} + 10 \log \frac{B_{rf}}{2B_v} + 9$$

where C/N is the carrier-to-noise ratio that was found earlier. The last term of 9 dB results from converting from rms to peak-to-peak signal values, and the next-to-last term results from the bandwidth considerations.

In the preceding example, we found that the carrier-to-noise ratio of a particular receiver was 58.2 dB. In that example, we assumed that the bandwidth of the rf portion of the receiver was 15 MHz. Let us further assume that the video bandwidth of the receiver is about 4.5 MHz. The equation for the signal-to-noise ratio now becomes:

$$\frac{S}{N} = 58.2 + 10 \log \frac{15}{2(4.5)} + 9$$
$$= 58.2 + 2.2 + 9$$
$$= 69.4 \text{ dB}$$

Thus, the actual signal-to-noise ratio at the output of the receiver is 69.4 dB.

The above calculations were made assuming ideal conditions. It was assumed that the power of the transmitter is actually all radiated. In practice, there are losses in the transmission lines for both the transmitting and receiving antennas, and these losses must be taken into consideration.

A more important factor that must be taken into consideration in any microwave link is the fact that microwave signals are subject to fading due to atmospheric conditions. As pointed out earlier, fading is unpredictable, and most microwave links are rated statistically. Usually the rating is in terms of the percentage of the time that the signal will fade to the noise level. Naturally, the better the signal-to-noise ratio under ideal conditions, the less likely it is that fading will seriously impair the signal quality.

FM RECEIVER PERFORMANCE

There are two factors that must be taken into consideration when frequency modulation is compared with amplitude modulation. One is that fm is somewhat less susceptible to noise than a-m; this is accounted for by the fm improvement factor which we will discuss shortly. The other is that the threshold of an fm receiver is somewhat higher than the actual noise level in the receiver.

Noise in a receiver tends to consist of spikes. The peak value of these spikes is much higher than the rms value. When an fm receiver is receiving two signals, it tends to lock onto, or *capture*, the larger of the two signals. Thus, in the presence of noise and a weak signal, the receiver will tend to capture the noise spikes rather than the signal, even though the rms value of the signal is higher than the rms value of the noise. As a result of this phenomenon, an fm receiver will tend to capture the noise rather than the signal until the rms value of the signal is 9 or 10 dB greater than the rms value of the noise. Once this threshold is passed, there is actually an improvement from using fm under the proper conditions.

The fm improvement factor is given by:

$$10 \log 3 \left(\frac{D}{B_v} \right)^2$$

where,

B_v is the video bandwidth of the receiver,
D is the deviation of the fm transmitter.

Note that B_v and D must be expressed in the same units, usually megahertz.

In most tv microwave links, the deviation is not greater than the maximum video frequency to be transmitted, because the tradeoff between improvement of the noise performance is not worth the additional spectrum space required. Thus, the improvement factor will be about:

$$10 \log 3 = 4.8 \text{ dB}$$

There is still another factor in fm transmission that may be used to improve the performance of the system, particularly with color signals. This is the addition of pre-emphasis at the transmitter and a corresponding amount of de-emphasis at the receiver. A pre-emphasis network is actually a high-pass filter that will artificially increase the level of the high-frequency portion of the video signal. Inasmuch as the color information is transmitted about 3.58 MHz above the bottom end of the spectrum of the video signal, the color information in the signal will be emphasized. At the output of the receiver, after the signal is no longer affected significantly by the noise in the system, the signal is passed through a low-pass filter that will restore the proper level of all components of the signal. The improvement commonly obtained by pre-emphasis and de-emphasis is about 2 dB.

Going back to the example that we have been using, if the receiver used fm instead of a-m, we would expect the signal-to-noise ratio to be about:

$$\frac{S}{N} = 69.4 + 4.8 + 2 = 76.2 \text{ dB}$$

This shows that a definite improvement can be obtained by using fm. It should be noted, however, that the improvement depends on the deviation of the fm transmitter. If a narrow-band fm system, with a small amount of deviation, is used, the fm improvement factor can become negative, and the performance will be inferior to that of an a-m system.

HETERODYNE MICROWAVE SYSTEMS

In addition to the usual a-m and fm microwave systems, there is another form of transmission that is becoming increasingly popular. In this system, shown in Fig. 13-8, there is no modulator in the usual sense of the word. The input signals occupy the same portion of the spectrum that they would in a cable tv system; that is, they are in the 54- to 300-MHz region. Signals of this type may be fed to the system from the output of a conventional headend. The signals are transformed to the microwave region by heterodyne conversion. The heterodyne converter is usually called an *up converter*.

In the heterodyne process, the output of the converter consists of three frequencies. These are the local-oscillator frequency, and the sum and difference of the signal frequency and the local-oscillator frequency (Fig. 13-9). When the conversion is from the vhf region

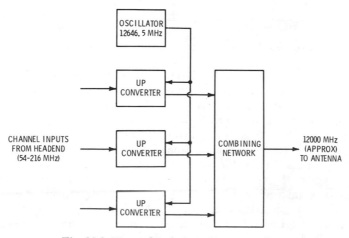

Fig. 13-8. Heterodyne microwave transmitter.

of the spectrum to the microwave region, there is such a great difference in frequency that it is quite easy to filter out any unwanted components. This is exactly what is done in the heterodyne system. For example, if the input signal happens to be channel 2, which occupies the range from 54 to 60 MHz, and the local-oscillator signal has a frequency of 12,646.5 MHz, the output of one up converter will contain the following frequencies:

$$f_0 = 12646.5$$
$$f_1 = 12646.5 - (54 \text{ to } 60) = 12586.5 \text{ to } 12592.5 \text{ MHz}$$
$$f_2 = 12646.5 + (54 \text{ to } 60) = 12700.5 \text{ to } 12706.5 \text{ MHz}$$

All of these frequencies except the upper sideband (f_2) are filtered out in the up converter. Thus, the microwave signals are exactly the same as the original signals except that they are shifted to the microwave portion of the spectrum.

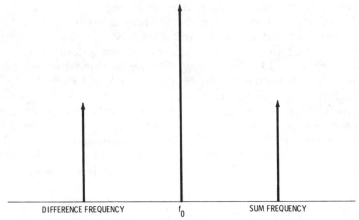

DIFFERENCE FREQUENCY f_0 SUM FREQUENCY

Fig. 13-9. Frequencies generated in heterodyne process.

In the transmitter, each channel is processed separately, and after being shifted to the microwave range the signals are then combined in a passive network and fed to one or more antennas. In the general case, there will be a separate antenna for each direction in which it is desired to transmit the signals. The separate processing of each channel minimizes any cross modulation that might otherwise occur due to the nonlinearities of the modulator.

Fig. 13-10 shows a block diagram of a receiver used with a heterodyne microwave system. In general, it is the inverse of the transmitter. Signals are picked up at the microwave frequency and are down-converted to their original place in the spectrum. An interesting feature of some heterodyne systems is that the local oscillator of the receiver can be phase-locked to the local oscillator in the transmitter. This is accomplished by transmitting a pilot tone in the blank portion of the spectrum between channels 4 and 5. When the local oscillators in the transmitter and receiver are phase-locked, the effect is the same as though the signals coming out of the receiver were phase-locked to the original television transmitter. This avoids problems that might otherwise be encountered because of frequency shifts in the system. It also relieves the requirement for an extremely stable local oscillator in the receiver.

In the receiver, all channels are processed in a single down converter. This does not cause cross modulation problems because the signals have a very low level.

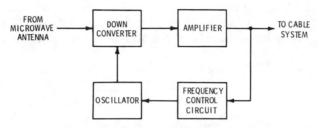

Fig. 13-10. Heterodyne microwave receiver.

SUPERTRUNK

We have shown repeatedly that the losses of a coaxial cable are relatively high at normal television frequencies. We also pointed out in an earlier chapter that the loss in a coaxial cable increases approximately as the square root of frequency. The only solid reason for cable tv systems to operate at the same frequencies as television broadcast stations is that the subscribers' receivers operate at these frequencies. The use of any other frequency would require an expensive converter at each subscriber's receiver.

In all of the above justification for operating a cable tv system on the normal tv channels, there is nothing to indicate that it might not be economically feasible to use lower frequencies if the cable were used only between two points such as a remote headend and a local distribution point. In fact, cable is used in this way. In order to minimize the loss in these so-called *supertrunk* cables, the lowest practical frequencies—usually only a few megahertz—are often used.

A typical supertrunk system might carry as many as seven tv channels in the frequency range of 6 to 48 MHz. If a 0.75-inch coaxial cable is used at these frequencies, the spacing between amplifiers may be as much as 1.5 miles compared with about 0.4 mile

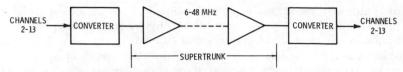

Fig. 13-11. Supertrunk distribution system.

for the regular tv channel frequencies. Runs as long as 20 miles are possible with enough signal-to-noise ratio left to feed a regular cable distribution system.

Fig. 13-11 shows a block diagram of a supertrunk system carrying signals from a remote site to a local distribution hub. At the head-end, all of the signals are converted to the frequency range between 6 and 48 MHz. All of the amplifiers along the supertrunk operate over this frequency range. It is worth noting that amplifier design at these lower frequencies is much simpler than at higher frequen-

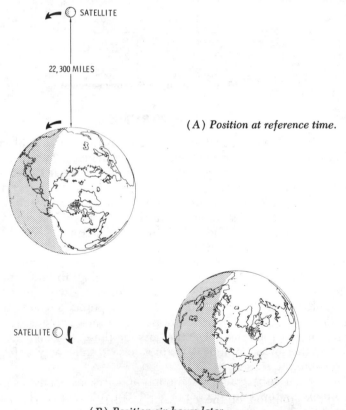

(A) Position at reference time.

(B) Position six hours later.

Fig. 13-12. Satellite in synchronous orbit.

cies. At the receiving end, all of the signals are converted back to the regular vhf channels.

In some cable tv systems, the supertrunk concept is used for regular signal distribution through the system. All of the signals in the main trunk of the system are at the lower frequencies. At distribution points throughout the system, the signals are converted back to the tv channels in order to be compatible with the subscribers' receivers.

SATELLITES

A subject that no cable tv system operator can afford to neglect is the transmission of tv signals from orbiting satellites. The use of such satellites as a medium for getting signals to cable systems is only a matter of economics. They are already being used in tv broadcasting.

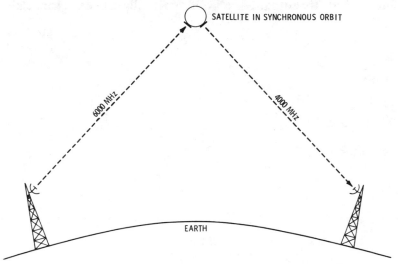

Fig. 13-13. Satellite transmission system.

A communications satellite orbiting the earth is launched into a synchronous orbit so that it does not move with respect to a given spot on the surface of the earth. Fig. 13-12 (not to scale) shows a satellite in synchronous orbit around the earth. When the satellite is approximately 22,300 miles above the earth, it will orbit the earth at a velocity of about 7000 miles per hour. This velocity matches the speed of rotation of the earth. Fig. 13-12A shows the earth and satellite as seen from above the north pole of the earth. In Fig. 13-12B, which shows the situation 6 hours later, both the earth and the satellite have rotated through 90 degrees, and the satellite is still over the same point on the surface of the earth.

The use of a synchronous orbit means that the receiving and transmitting antennas on the surface of the earth can be pointed in one direction and left there. It is not necessary for the antennas to track the satellite as it would be in the case of a space vehicle that was not in synchronous orbit.

Fig. 13-13 shows a simplified diagram of a satellite system. At the transmitting end, the tv signals are frequency-modulated onto a 6000-MHz carrier and transmitted to the satellite. At the satellite, the signals are received and are down-converted to about 4000 MHz. A typical system of the type in use today can carry up to 12 tv channels between 3700 and 4200 MHz.

Frequency modulation is used to improve the signal-to-noise ratio and to reduce the power requirements on board the satellite. Each channel occupies as much as 40 MHz of bandwidth. This includes guard bands that are used to avoid cross talk between channels.

CHAPTER 14

Cable TV Instrumentation and Test Equipment

The instrumentation and test equipment used for maintenance and repair of cable tv systems and components includes many of the instruments that would be used in the maintenance of just about any type of electronic equipment. A good multimeter is essential. An oscilloscope is also essential, particularly on the bench, and it is often very useful in the field. In addition to these conventional instruments, there are other types of specialized test equipment used in cable tv. The average technician may not be familiar with this specialized test equipment, so some of the most commonly used instruments will be described in this chapter.

SIGNAL-LEVEL METER

The *signal-level meter,* or slm, is actually a tuned radio-frequency voltmeter. This instrument is also known as a *field-intensity meter,* or fim, or as a *field-strength meter,* or fsm. Because of the fact that it is so useful in checking the levels of signals at the subscribers' terminals, it is frequently called an *installer's meter.*

Regardless of the name applied to it, the slm is a voltmeter, and like any other voltmeter, it will respond to the voltage applied to its terminals. In addition to being calibrated in terms of voltage, most slm's are also calibrated in dBmV. The input impedance of the slm is usually 75 ohms, so the dBmV indication will be accurate when the meter is connected directly to a 75-ohm cable system.

Signal-level meters are used with dipole antennas to measure the strength of signals in the air rather than on the cable. A calibration curve is provided so that the field intensity in microvolts per meter can readily be determined. This application of the slm is useful in finding sources of interference or checking rf leakage from the cable system.

Although there are many sophisticated instruments available for cable tv work, the slm is one of the most versatile instruments that the cable tv technician can have. Most slm's are portable so that they can be carried to any point in the system. Because the slm is tunable, it can be used to measure the levels of the visual and aural carriers separately to set the traps in various parts of the system. In addition to tests that can be made throughout the system, the slm can also be used on the bench for checking amplifiers and other components.

The main precaution to be observed in using an slm is to be sure that it is connected properly to the source of the signal to be measured. Although the slm will respond to the voltage that appears at its input, this may not be the voltage that you are trying to measure unless the slm is connected properly. If the slm is connected to the system through 75-ohm coaxial cable with good connectors, the indication will be reliable. If it is connected in haywire fashion, the indication may be meaningless.

Fig. 14-1 shows an equivalent circuit of an slm connected to a 270-MHz signal source through a 75-ohm transmission line. The effective impedances of both the slm and the signal source are each 75 ohms, so the 10 volts that is available from the source will divide across the two impedances. The voltage across the input of the slm, and consequently its indication, will be 5 volts. This is a good indication because it is the voltage that would appear across any 75-ohm load connected at this point. This is usually what we are interested in measuring in a cable tv system.

Fig. 14-2A shows the equivalent circuit of the same slm connected to the same signal source through a pair of "haywire" leads that, for convenience, we will assume to have a characteristic impedance of 300 ohms, which is a reasonable value. To illustrate just how

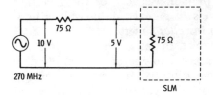

Fig. 14-1. Equivalent circuit of an slm connected to a 75-ohm source.

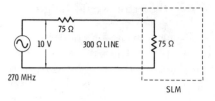

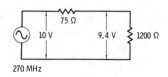

(A) Equivalent circuit.　　　　(B) Source terminal voltage.

Fig. 14-2. Slm connected to 75-ohm source through quarter-wave 300-ohm line.

much trouble haywire leads can cause, assume that the leads are a quarter wavelength long, which is only a few inches. The impedance seen looking into a quarter-wavelength transmission line is given by:

$$Z_{in} = \frac{Z_o^2}{Z_L}$$

where,

Z_{in} is the impedance seen looking into the line,
Z_L is the load impedance connected to the line,
Z_o is the characteristic impedance of the line.

In our example, the impedance seen by the source looking into the haywire leads is:

$$Z_{in} = \frac{(300)^2}{75} = 1200 \text{ ohms}$$

This means that we have the equivalent circuit shown in Fig. 14-2B. Here the voltage that will appear at the point where the leads are connected to the source will be 9.4 volts. We still have not determined what the actual indication of the slm will be. We can get some idea of this by assuming that there is no loss or radiation from the haywire leads and that all of the power delivered by the source reaches the slm. This probably is not true, but it will help to give some idea of how the indication can be affected.

The power delivered by the source to the leads will be:

$$P = \frac{E^2}{R} = \frac{(9.4)^2}{1200} = 0.07 \text{ watt}$$

If all of this power reaches the slm, the voltage at its terminals will be given by:

$$E = \sqrt{PR} = \sqrt{0.07 \times 75} = 2.3 \text{ volts}$$

Thus, our haywire arrangement caused the slm to indicate 2.3 volts as compared with the 5 volts that we would obtain if the instrument were connected properly. This is an error of about

54%. If we were using the dBmV scales, the error would be greater than 6 dB.

The above example shows that the most important consideration in the use of an slm is that it be connected to the source in a way that does not disturb the voltage being measured. The best way is to make connections through 75-ohm cable whenever possible. Even a few inches of random wire leads can cause serious errors in the indication.

SWEPT-FREQUENCY OSCILLATORS

The *swept-frequency oscillator*, also known as a *sweep generator*, a *sweeper*, or simply a *sweep*, is an oscillator that automatically sweeps a signal over a predetermined frequency range. Fig. 14-3 shows an instrument of this type that is designed to cover frequencies of interest in cable tv work.

Fig. 14-3. Sweep generator.

The value of a swept-frequency signal source can be realized by considering the work that must be done when a conventional, manually tuned signal generator is used to determine the frequency response of a device. Fig. 14-4 shows a test setup used to measure the frequency response of a bandpass filter. In order to determine the frequency response, the signal generator is first tuned to the

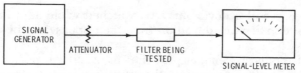

Fig. 14-4. Measurement of frequency response with signal generator and
signal-level meter.

lowest frequency of interest, and the output is set to some predetermined level. The output of the filter is read on the meter and
recorded. The signal generator is then set to the next higher frequency, and the process is repeated. After all of the necessary
measurements have been made, the results are plotted as shown in
Fig. 14-5. To be sure that all of the little variations in the response
curve are detected, measurements must be made at closely spaced
frequencies. This is time-consuming, particularly when the results
show that the frequency response is incorrect and adjustments must
be made. When this happens, the entire process must be repeated
to ensure that the response is correct.

Fig. 14-6 shows the setup required when a sweeper is used to
make the same frequency-response measurement. The start and
stop frequencies of the oscillator may be set by means of dials on
the front panel. Internal circuitry ensures that the level of the output
frequency will remain constant over the entire frequency range. The
detector is usually contained in the sweeper, although an external
detector may be used.

The output of the detector is connected to the vertical-deflection
circuits of a cathode-ray oscilloscope. The horizontal input of the
oscilloscope is fed by a sawtooth wave from the sweeper. The sawtooth wave is such that when the sweeper is at its lowest frequency,
the trace will be at the left side of the crt. Thus, as the output of
the sweeper goes from its lowest to its highest frequency, the trace
will move from left to right across the face of the crt. The amount of
vertical deflection depends on the output of the filter being tested.
Thus, the trace on the crt will be a plot of the response of the filter
as a function of frequency, similar to that shown in Fig. 14-5.

The amount of time that can be saved by using a sweeper for
frequency-response measurements is obvious. The frequency-

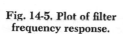

Fig. 14-5. Plot of filter
frequency response.

249

response curve of a device may be watched while adjustments are being made so that any deviation from the response can be corrected quickly.

One of the most common problems in using sweepers is that the sweep rate is often set so fast that the circuit under test does not have time to respond while the signal is sweeping through its pass-

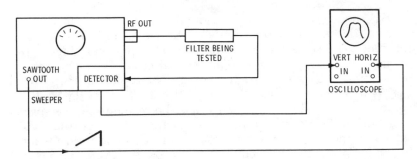

Fig. 14-6. Measurement of frequency response with sweeper and oscilloscope.

band. Fig. 14-7 shows two different displays that might be obtained when a sweeper is used to measure the frequency response of a cable tv trap. The only difference in the two measurements is the time required to sweep the display from the left of the screen to the right. Obviously, in Fig. 14-7A, the sweep is so fast that the trap does not have time to respond. This is usually called *scan loss* and must be carefully avoided if we are to get good data from a test. If there is any question that scan loss may be present, the sweep speed should be reduced.

Several accessories are available for use with sweepers. One of the problems with using a swept frequency is the difficulty in obtaining accurate identification of the frequency at any point in the display. This process is aided by the use of *marker generators* that generate distinctive marks on the display at particular frequencies. Both fixed-frequency and variable-frequency marker generators are available.

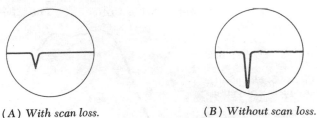

(A) *With scan loss.*　　　　　(B) *Without scan loss.*

Fig. 14-7. Oscilloscope displays for fast and slow sweeps.

250

Other accessories include bridges that can be used instead of detectors. A bridge may be used with a sweeper in the arrangement shown in Fig. 14-8. Here the bridge provides an output that is proportional either to the reflection coefficient of the device under test or to its return loss. Thus, the display on the oscilloscope will show either the reflection coefficient or the return loss of the input port of the device being tested, as a function of frequency.

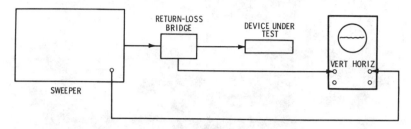

Fig. 14-8. Use of sweeper to measure return loss.

THE SPECTRUM ANALYZER

Any technician who has ever used a cathode-ray oscilloscope knows that there is no substitute for seeing a display of a signal. The graphic display of the signal gives more information in a shorter period of time than any other measurement. There are many places in a cable tv system where an oscilloscope can be used to advantage. However, one place where the oscilloscope is practically worthless is in looking at the signals that appear on the cable of a system that is carrying several channels. In the first place, an oscilloscope that has a frequency response up to 300 MHz is expensive. More important, there are only two conductors in a cable, so the voltage that appears at any point along the cable is the algebraic sum of all the signals that are being carried in the system. Even if we had an oscilloscope that had adequate frequency response, we could not make the display stand still on the screen. The result would be a meaningless collection of moving lines.

There is a way that we can view the signal on a cable system on the screen of a cathode-ray tube, and that is by means of an instrument called a *spectrum analyzer*. Fig. 14-9 shows a spectrum analyzer designed specifically for cable tv work.

The vertical deflection on the cathode-ray tube of a spectrum analyzer is a function of the applied voltage, and the horizontal deflection is a function of the frequency of the applied signal. Fig. 14-10 shows what the display of a spectrum analyzer would look like if it were connected to a cable carrying a single tv signal. It can be seen that the amplitudes of the various signals are displayed

251

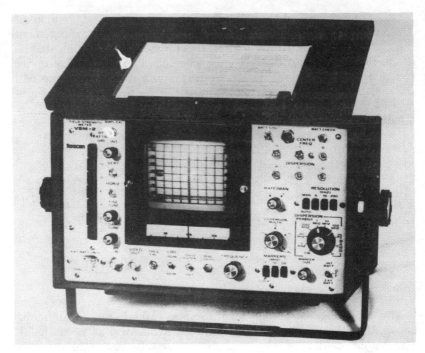

Fig. 14-9. Cable-tv spectrum analyzer.

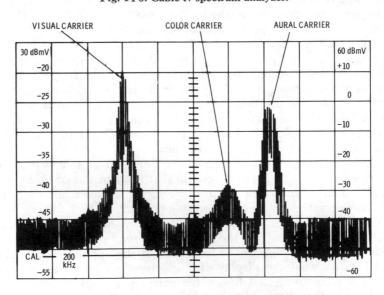

Fig. 14-10. Single-channel display on a spectrum analyzer.

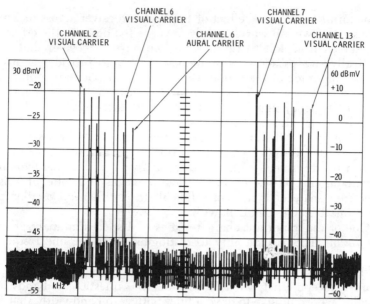

Fig. 14-11. Twelve-channel display on a spectrum analyzer.

as a function of frequency. The visual carrier, the color subcarrier, and the aural carrier are clearly visible. Fig. 14-11 represents a spectrum-analyzer display of the vhf channels on a cable tv system. Note that because a wider range of frequencies is displayed on the screen of the crt in this case, the details of the signals are not as easily distinguished.

Fig. 14-12 shows a block diagram of a spectrum analyzer. It is similar to a superheterodyne receiver. The i-f amplifier contains a very narrow-band filter that (ideally) allows only one frequency to pass through the i-f amplifier at any one time. The local oscillator of the spectrum analyzer is swept across its range by a sawtooth

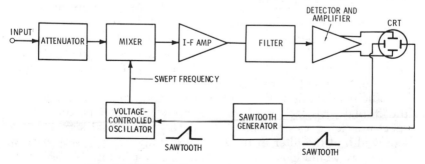

Fig. 14-12. Block diagram of spectrum analyzer.

253

waveform. This has the effect of tuning the receiver across its range. The same sawtooth waveform also sweeps the beam of the crt across the screen from left to right. Thus, we have a receiver that automatically sweeps across a frequency range and provides a display of the amplitude of the signals that it picks up. The horizontal position of each signal on the screen depends on its frequency; the lowest-frequency signals are at the left of the screen, and the highest-frequency signals are at the right.

Just as two different settings of the controls of an oscilloscope will provide two different patterns from the same signal, the display shown on the screen of a spectrum analyzer will depend to some extent on the setting of the various controls. In spite of this, a spectrum analyzer is no more difficult to use than an oscilloscope, once one becomes familiar with it.

One of the important characteristics of a spectrum analyzer is its *resolution*. This is its ability to distinguish between one frequency and another. The resolution is determined by the bandwidth of the filter in the spectrum analyzer and the speed with which the spectrum is swept. First, we will consider the question of bandwidth. Ideally, we would like to have a filter with zero bandwidth, that is, a filter that would pass only one frequency. Of course, it is impossible to build a filter with zero bandwidth, and even if we could build such a filter, it would have an infinitely long time delay, so we would never get any output from it. The response time of any filter is a function of its bandwidth. The narrower the bandwidth of a filter, the longer the time it takes to provide an output after an input is applied to it. Thus, the bandwidth that we can use in the filter of a spectrum analyzer is limited to some extent by the speed at which we wish to sweep across the spectrum.

In the interest of eye comfort, we want to sweep the spectrum analyzer across the spectrum at a speed that will not cause objectionable flicker of the display. The same general considerations that apply to flicker in a tv picture apply to a spectrum analyzer. Of course, more flicker can be tolerated in the display of an instrument than can be tolerated in a picture that will be viewed for a longer period of time. Usually the slowest rate at which we want to sweep a spectrum is about 30 times per second, and a 60-sweeps-per-second rate is much more comfortable to view. Thus, the bandwidth of the filter cannot be too narrow.

If the bandwidth of the filter in a spectrum analyzer is too wide, the display will lack resolution. Suppose that the portion of the spectrum being viewed contains only one sinusoidal signal. If the bandwidth of the filter in the spectrum analyzer were very narrow, the display would look very much like the plot in Fig. 14-13A. If the bandwidth is too great, the display will look like that shown in Fig.

14-13B. This display is actually the frequency-response curve of the filter in the analyzer. This is not suprising because what we are doing is sweeping a sinusoidal signal in frequency, just as we would do to get a display of the frequency response of a filter.

This brings us to the question of how sweep speed affects the display of a spectrum analyzer. We noted earlier that any filter takes some time to provide an output after an input is applied to it. This, of course, is true of the filter in a spectrum analyzer. The simple rule is that the signal being displayed must be within the passband of the filter long enough for the output to reach its full value. When the frequency is swept too fast, the display will be distorted.

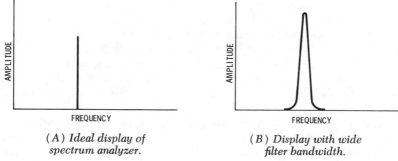

(A) Ideal display of
spectrum analyzer.

(B) Display with wide
filter bandwidth.

Fig. 14-13. Effect of filter bandwidth on spectrum-analyzer display.

The smallest frequency increment that a spectrum analyzer can resolve is usually called its *resolution,* or *resolution bandwidth.* It is usually specified in terms of the frequency range between the points at which the response of the filter is down 3 dB. Thus, a spectrum analyzer designed for cable tv use might have resolutions of 500 Hz, 10 kHz, and 200 kHz. The smallest resolution would be used for looking at the spectrum of video signals, whereas the largest resolution would be used for looking at the entire spectrum from dc to 300 MHz.

The range of frequencies that is displayed on the screen at any time is usually called the *dispersion* of the analyzer. The dispersion is usually specified in terms of the amount of frequency covered in one division on the screen. Thus, a typical specification for dispersion might be 3, 15, and 200 kHz and 1, 10, and 35 MHz per scale division on the screen of the analyzer. In addition, many spectrum analyzers have a continuously adjustable dispersion.

The vertical scale of the spectrum analyzer may be either linear or logarithmic. When linear deflection is used, the vertical deflection is simply proportional to voltage. By means of the input attenuator, the scale may be calibrated directly in microvolts or millivolts. With a logarithmic scale, the vertical deflection may be calibrated in dBmV.

SYNCHRONOUS OR SUMMATION SWEEPING

Earlier in this chapter, we described both swept-frequency sources and spectrum analyzers. Each of these instruments can be used to check either components or complete cable tv systems. It is also possible to use the two instruments together to perform tests on a cable tv system while the system is in operation, with a minimum amount of interference to the subscriber's reception. This type of test is usually called *synchronous* or *summation* sweeping. The test setup is shown in Fig. 14-14.

At the input to the cable system at the headend, a signal from a sweeper, or transmitter, is connected to the system through a directional coupler. The signal level is adjusted so that it is higher than the level of any of the tv signals that are being carried on the cable. At the point along the system where a measurement is to be made, a receiver (which is actually a spectrum analyzer) is connected, usually through a directional coupler.

The signal from the transmitter sweeps through all or part of the frequency band carried by the cable system. The receiver detects the signal from the transmitter and locks onto it. Thus, the receiver operates like a spectrum analyzer sweeping through the spectrum in step with the signal from the transmitter. The receiver usually has a built-in filter that rejects the horizontal sweep rate of the tv signals, thus minimizing interference from the signals being carried on the system.

There is little interference to the subscriber's reception because the swept signal is not in any one tv channel for more than about

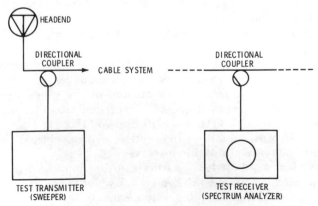

Fig. 14-14. Setup for synchronous sweeping of cable tv system.

50 microseconds. This means that the test signal could interfere with any of the pictures carried on the system for only a few lines at the most. Some arrangements repeat the sweeping long enough to get a plot of the system response. In other arrangements, the receiver has a long-persistence screen on the cathode-ray tube so that a single sweep is all that is required to determine the response of the system. The display on the receiver can be photographed with the type of camera that is normally used to photograph oscilloscope displays. Photos of the response of various parts of the system can be kept on file so that they will show how the system response looks when the system is operating properly. This makes it easy to tell when there is trouble in some part of the system.

In general, the synchronized sweep system works best when the receiver is connected to the system at a point where there is no tilt. If it is desired to sample the signal at a point in the system where the tilt is appreciable, a length of cable can be inserted in series with the receiver input to get a flat response.

This method of testing systems can be applied at any point in the system. For example, the response can be checked both before and after an amplifier or a section of the cable system. The synchronized-sweep method of testing can serve as the basis of an automatic monitoring system for a cable tv system.

WAVEFORM MONITORS

An instrument that has been used for many years in tv broadcasting and is now finding increased use in cable tv systems is the waveform monitor (Fig. 14-15). The waveform monitor is an oscilloscope, but it has internal circuits that suit it to the observation of tv waveforms. The input is a video signal. This may be at the headend between the demodulator and the modulator, in the origination studio, or from a demodulator that is attached along the

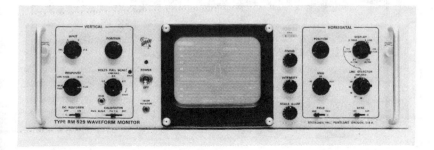

Courtesy Tektronix, Inc.

Fig. 14-15. A tv waveform monitor.

cable. As cable tv systems become more automated, this last application will become more significant.

The waveform monitor has controls on the front panel that allow displaying one or two fields of the tv signal or any one or two of several lines of the picture. Television stations transmit test and reference signals on some of the lines of the picture that occur during the vertical-blanking interval. Thus, these signals normally are not visible on the receiver screen.

Test and reference signals transmitted by the tv station are used for remote control of the transmitter and for automation of the station. These signals can also be used to provide a great deal of information about how a cable tv system is operating.

Test and reference signals will become more important with increasing automation of cable tv systems. Although with a little ingenuity you can view the test signals on an ordinary oscilloscope, the process is simplified considerably when a waveform mointor is used for the purpose.

THE PICTURE MONITOR

An instrument that is indispensable in a tv broadcast station and is becoming more popular in cable tv systems is the picture monitor (Fig. 14-16). The picture monitor is really the video and sync portions of a tv receiver, but it usually has much tighter specifications than an ordinary tv receiver. Provisions are made for viewing either video or rf signals.

Courtesy Tektronix, Inc.

Fig. 14-16. A tv picture monitor.

258

Picture monitors have additional features that permit more detailed examination of the picture than is possible on a tv set. One such feature is an expanded vertical scale that permits observation of the details of the vertical-blanking pulse.

Two places where picture monitors are very useful in a cable tv system are at the headend and in the origination studio. At the headend, the picture monitor can be connected after a demodulator so that the operator can judge the quality of the picture that he is picking up from a tv station. Remember that the picture quality will be best at the headend and will only become degraded as the signal passes through the system. The picture on the subscriber's receiver can never be quite as good as the picture seen on the monitor at the headend.

In the studio, the picture monitor is used to control the quality of the picture being produced. Although all of the information about picture quality is available on the waveform monitor, there is no substitute for actually looking at the picture itself for getting an idea of its quality.

THE TIME-DOMAIN REFLECTOMETER

The *time-domain reflectometer*, or *tdr* as it is commonly called, operates directly on the principles of transmission-line theory. In an earlier chapter, we discussed the fact that if a cable were not terminated in its characteristic impedance, there would be a reflection from the end. The tdr sends an impulse down the cable and then "listens" for a reflection.

A block diagram of a tdr is shown in Fig. 14-17. A step waveform is fed to the line being tested by some sort of diplexer such as a bridging tee. The impulse travels down the line. At the same time, a sawtooth voltage is applied to the horizontal channel of the oscilloscope. Reflections from any part of the line are coupled to the

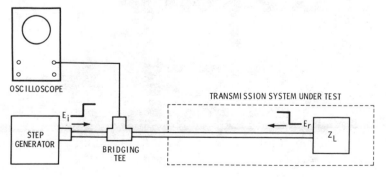

Fig. 14-17. Block diagram of time-domain reflectometer.

vertical-deflection channel of the oscilloscope. The display on the oscilloscope is as shown in Fig. 14-18. Here the vertical pip indicates a reflection, and the distance along the horizontal axis indicates the distance to the discontinuity on the cable that caused the reflection. The horizontal deflection may be calibrated either in feet or in meters.

(A) *Properly terminated line.*   (B) *Improperly terminated line.*

Fig. 14-18. Displays on time-domain reflectometer.

When the time-domain reflectometer is applied to the input of a properly terminated section of good cable, there will be practically no reflection at all. However, if there is trouble on a cable such as might be caused by water leaking in, there will be a reflection, and by reading the horizontal distance on the screen, the technician can find the distance to the fault to within a foot or so.

Some reflectometers, such as the one shown in Fig. 14-19, use a strip chart instead of an oscilloscope display. This arrangement is handy because a record can be kept of the reflection characteristics of each portion of the system. If trouble develops, the measurement can be repeated and the results compared with a record made when the system was operating properly.

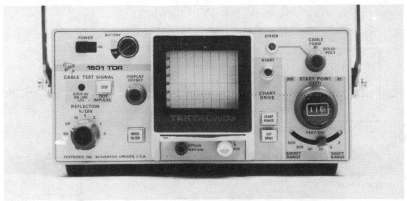

Fig. 14-19. A time-domain reflectometer with chart recorder.

THE TUNED FREQUENCY COUNTER

Although frequency counters are becoming increasingly common for frequency measurements in all types of electronic equipment, these instruments are of limited value in making frequency measurements on cable tv systems. The reason is that the front end of a conventional frequency counter is "wide open." It is designed to count unknown frequencies, so there is no point in restricting the bandwidth at its input. When such an instrument is used to measure frequency on a cable tv system, it will try to count the visual carrier, the video modulation, and the aural carrier. The result is that the measurement is meaningless.

In the next chapter, we will discuss ways in which a conventional counter can be used to make frequency measurements on a cable tv system, but at this point we should mention that special counters are available that will work on cable tv systems without additional equipment. These special counters are called *tuned frequency counters*. As the name implies, the input circuits of a tuned frequency counter are tuned so that only a narrow band of frequencies will be available at the counter circuits. In addition, a limiter is provided to remove the modulation from the visual carrier.

CHAPTER 15

Proof-of-Performance
and
System Measurements

The term "proof of performance" originated with the FCC in connection with radio and television broadcasting. The FCC insists that proof-of-performance measurements be made to ensure that the station is operating within the rules and the parameters set forth on the station license. When the FCC began to regulate cable tv, it naturally used the same terminology.

The parameters of a cable tv system that the FCC has regulated are listed in Table 15-1. Because the FCC Rules and Regulations are being continually updated, particularly as a result of the work of the Cable Television Advisory Committee, the limits on the parameters listed in Table 15-1 are subject to change. The operator of a cable tv system should be familiar with the latest FCC requirements.

The fact that the FCC requires proof-of-performance measurements before issuing a Certificate of Compliance is not the only justification for system measurements. Certainly, failure to comply with the FCC standards will lead to sanctions, but more important, failure to deliver high-quality pictures will result in loss of subscribers. System measurements should be made on a periodic basis to ensure that the operator is getting the most from his system.

In this chapter we will discuss methods of measuring the parameters listed in Table 15-1. Although emphasis will be placed on the advantages of using the best test equipment available, consideration will also be given to the fact that many small cable tv systems

Table 15-1. FCC Proof-of-Performance Testing Requirements

Test	Specification
Visual carrier frequency	± 25 kHz
Visual carrier frequency at output of set-top converter	± 250 kHz
Aural carrier frequency	4.5 MHz ± 1 kHz above visual carrier frequency
Visual carrier level	0 dBmV minimum Maximum not to cause receiver overloading
Permissible variation of visual carrier level A. Any channel over a 24-hr period B. Between any two adjacent channels C. Between any two channels	 12 dB maximum 3 dB maximum 12 dB maximum
Hum level	± 5% of visual carrier level maximum
Frequency response (each channel)	± 2 dB from 0.75 MHz below to 4 MHz above visual carrier frequency
Aural signal level	13 to 17 dB below visual carrier level
Signal-to-noise ratio for all signals picked up or delivered within their Grade-B contour	36-dB signal to noise minimum 36-dB signal to co-channel minimum
Ratio of signal to intermodulation and non-offset carrier	46 dB minimum
Isolation between subscriber's terminals	18 dB minimum (greater if required)
Radiation A. 54 MHz or below B. 54 MHz to 216 MHz C. Above 216 MHz	 Less than 15 μV/m at 100 ft Less than 20 μV/m at 10 ft Less than 15 μV/m at 100 ft

Notes:
1. Proof-of-performance measurements must be made on at least each Class-I channel carried by the system.
2. Proof-of-performance measurements must be made at least one each calendar year, with the period between measurements not greater than 14 months.
3. Measurements must be made at not less than three widely separated points in the system. At least one of the points shall be at the greatest distance from the system input in terms of cable length.

simply cannot afford elaborate test equipment. If such systems are to survive, they must be able to make the necessary measurements with a minimum amount of test equipment. In general, a technician with a great deal of ingenuity and a small amount of test equipment can bring a cable tv system to optimum operating condition. The price that he pays for not having elaborate test equipment is time. In fact, one of the greatest justifications for acquiring test equipment is to save technicians' time. This also saves money because the picture quality at the subscriber's terminal can be corrected more quickly if trouble is located quickly.

TEST-EQUIPMENT COMPATIBILITY

Before any meaningful measurements can be made, the test equipment must be connected to the cable system in such a way that it will not disturb the value of the parameter being measured. This means that connections should be made through coaxial cable and coaxial connectors. When a signal is taken from a cable, it must be taken through a tap or directional coupler so that the standing-wave ratio on the cable will not be increased.

The most common cause of incompatibility of test equipment with a cable tv system is that many test instruments which were not specifically designed for cable tv work, such as oscilloscopes and spectrum analyzers, have a 50-ohm input impedance instead of the 75-ohm impedance of almost all cable tv systems. When the input impedance of a piece of test equipment is 50 ohms, the instrument must be connected to the cable system through a pad that will make it look like a 75-ohm instrument to the system.

Fig. 15-1 shows a diagram of a pad that will transform the 50-ohm impedance of a test instrument to the required value of 75 ohms. This pad will introduce a loss of almost 6 dB, which is the minimum loss that will be introduced by a pad that accomplishes this particular impedance transformation. When such a pad is used, the value of 6 dB must be added to the indication of the test instrument.

Minimum-loss pads are commercially available, or they can be constructed by the technician. If a homemade pad is used, tests must be made to ensure that it behaves properly over the entire frequency range of interest. As with any other device connected to a

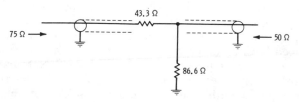

Fig. 15-1. Diagram of matching pad.

cable system, coaxial connectors must be used, and the device must be completely shielded.

Another instance of incompatibility of test equipment involves the fact that many instruments that cover the frequency range of interest, but were not designed specifically for cable tv use, are calibrated in dBm rather than in dBmV. When such an instrument is used, the reading must be corrected by adding 48.8 dB to the indication in dBm to get a value in dBmV. One dBm represents a power of 1 milliwatt across any value of impedance; one dBmV represents a voltage of 1 millivolt measured across a resistive impedance of 75 ohms. This is a power of only 1/75 microwatt. The value of 48.8 dB corrects for the difference in units.

Fig. 15-2 shows a 50-ohm spectrum analyzer connected to a cable tv system through a minimum-loss impedance-matching pad. The necessary corrections to get an indication in dBmV are shown in the figure.

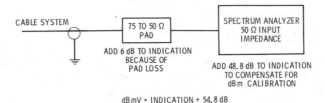

Fig. 15-2. Correction factors for 50-ohm instrument calibrated in dBm.

ACCURACY AND TRACEABILITY

When a technician buys a new instrument or uses an instrument that has proven to be reliable, he tends to think of the instrument in the same way he would think of a laboratory standard. He feels that the indication is always dependable. Of course, if instruments are properly cared for and their calibration is checked periodically, they will usually be very reliable. This, however, is a big "if."

Although the subject of calibration is beyond the scope of this book, every cable tv engineer and technician must have a way of satisfying himself that his instruments are operating properly and that their indications can be trusted. In standard measurements work, this subject is often called *traceability*. The fact that the calibration of an instrument is traceable means that its calibration can be traced back to a measurement standard. The usual standards of measurement are those maintained by the US National Bureau of Standards. This does not mean that every test instrument is actually compared with a measurement standard at the Bureau, but it does mean that the calibration of the instrument is periodically compared with the indication of an instrument that has been compared with a measurement standard.

Only the largest cable tv systems find it economically practical to maintain measurement standards. The usual practice is to send instruments out to a calibration laboratory periodically to have their calibration checked. Usually, laboratory or bench instruments hold their calibration better than instruments that are used in the field. Therefore, it is good practice to compare the instruments that are used in the field with bench instruments to ensure that the calibration has not changed.

One frequent type of error in measurements is the temperature-related error. Often, the temperature in a headend can vary over a wide range, and the temperature at which field measurements are made is certain to vary over a wide range. Any instrument that is to be used at widely differing temperatures should be calibrated not only at room temperature, but also at other temperatures, both high and low. With this information, the technician will know what type of error to expect when he is forced to make measurements at temperature extremes.

All measurements other than brief checks should be made only after the equipment has warmed up long enough that its indication will be stable.

In general, the accuracy of an instrument should be about ten times better than the tolerance of the quantity that is being measured. If instrument errors are large, a great deal of work is required to keep the system parameters within prescribed limits. Put another way, there is drift and inaccuracy enough in the system without adding more in the instruments.

GENERAL CONSIDERATIONS

The FCC rules are quite specific in stating that proof of performance measurements must be made under conditions that reflect the normal operation of the system. No changes can be made that will render the measurements easier to make if there is any chance that these changes will affect the performance of the system. Any microwave links that are normally in the system must be in operation when the proof measurements are being made. All of the amplifiers in the system must operate at their normal gain, even if it is necessary to set manual gain controls to this value or to insert a signal into the agc system to accomplish this. All pilot tones, substitute signals, and non-tv signals that are normally carried by the system must be carried while the measurements are being made.

Of course, there are some measurements, such as the noise on a channel, that simply cannot be made with simple test equipment while the signal is being carried. The FCC recognizes this and specifies that in such cases, the signal can be removed by disconnecting the antenna at the headend and substituting a matching resistor for the antenna. When this is done to make measurements on a particular channel, all of the other signals must be carried on the system.

The proof-of-performance measurements must be kept for at least five years, and they are subject to inspection at any time during this period. For this reason, not only must the measured values be retained, but also complete notes must be kept on how the measurements were made. This includes such things as the models and serial numbers of the instruments used, the dates that the instruments were last calibrated, and a diagram of the test setup. This type of information comes in very handy when one is trying to interpret or justify measurements that were made in the past by an engineer or technician who is no longer available.

FREQUENCY MEASUREMENTS

For the proof of performance of a cable tv system, the visual and aural carrier frequencies must be measured, and the frequency separation between them must be determined. Since most subscribers' tv sets use intercarrier sound, the frequency difference between the aural and visual carriers is probably the most important measurement from the point of view of subscriber satisfaction.

Because of the accuracy required, the best instrument to use for frequency measurements is the frequency counter. The biggest problem in measuring the frequency of the visual carrier is that the modulation tends to upset the measurement. For this reason, the counter cannot be connected directly to the cable system.

The easiest way to measure the visual and aural carrier frequencies is to use a tuned frequency counter. As explained in the preceding chapter, this instrument has a tuner on the front end and can tune separately to the visual and aural carriers. In addition, circuitry is provided to suppress the modulation on the carrier so that it does not interfere with the measurement.

Lacking a tuned frequency counter, the technician must find some other way to make the necessary frequency measurements. One technique that is sometimes used is to measure the frequencies of each of the oscillators in the headend and from these measurements calculate the frequencies of the visual and aural carriers of each channel. This technique can be used to get a good measurement of the frequencies, provided you are sure that you are not disturbing the frequency of the local oscillator when you attach the measurement device, such as a frequency counter.

Another method of measuring the visual and aural carrier frequencies is shown in Fig. 15-3. This method requires a signal source (which may be either a signal generator or a sweeper that is not in the sweep mode), two two-way splitters, a frequency counter, and a signal-level meter. Part of the signal from the signal source goes to the frequency counter, where the frequency of the signal source is measured. The other output of the splitter is fed to one of the *outputs* of the second splitter. This second splitter actually acts as a signal combiner. The cable is connected to the other output. The input connector of the splitter is connected to the signal-level meter.

Before the measurement can be made, the signal source must be tuned to approximately the frequency that is to be measured, and the level of the signal from the signal source must be adjusted to approximately the same level as the frequency being measured. This is done as follows. First, disconnect the signal source, and tune the signal-level meter to one of the frequencies to be measured. Note the indication of the signal-level meter. The tuning of the signal-level meter should not be changed during the rest of the measurement. Now disconnect the cable system and connect the signal source. Tune the signal source until an indication is obtained on the signal-level meter. Then adjust the level of the signal from the signal source until the signal-level meter indication is just about the same as it was when the meter was connected to the cable system.

Next, turn up the audio gain on the signal-level meter, and slowly tune the signal source until its output zero-beats with the frequency being measured. As the tuning approaches zero beat, a sharp whistle or tone will be heard. With a little practice, you can tune the signal source to zero beat rather easily. When the zero-beat condition is reached, the counter should be read and the indication logged.

When this method is used to measure an aural carrier, which has a lower level than a visual carrier, the output of the signal source must be reduced accordingly.

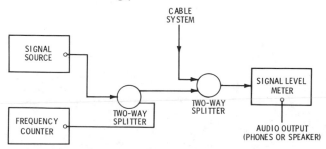

Fig. 15-3. Frequency measurement using audible beat from signal-level meter.

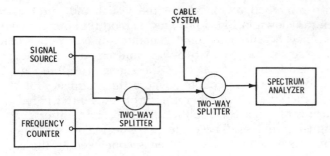

Fig. 15-4. Frequency measurement using visual beat displayed on a
spectrum analyzer.

A somewhat similar method of making frequency measurements
is shown in Fig. 15-4. Here the test setup is the same as in Fig. 15-3
except that a spectrum analyzer, rather than a signal-level meter, is
used to detect the zero-beat condition. The display will be as shown
in Fig. 15-5; the visual and aural carriers and the signal from the
signal source will be visible on the screen of the spectrum analyzer.
Slowly tune the signal source until it zero-beats with the carrier
whose frequency is to be measured. At zero beat, there will be a
noticeable jumping up and down of the signal. In general, this
method is much faster than the method shown in Fig. 15-3 because
you can see the signal from the signal source and know just how
much to change it to obtain zero beat.

Still another method of making frequency measurements is shown
in Fig. 15-6. This is similar to using a tuned frequency counter. An
instrument called a *signal processor* is used to select the carrier to
be measured and remove the modulation. The signal processor has
three outputs: the visual carrier, the intercarrier frequency, and
the detected video.

Usually, the intercarrier frequency is found by taking the difference between the visual and aural carriers. When using this

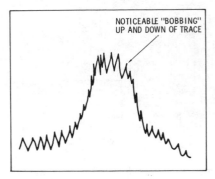

Fig. 15-5. Spectrum-analyzer display
near zero beat.

method, you should watch out for errors in what mathematicians call "the difference between two large numbers." Suppose that you have two frequencies, 200 MHz and 204.5 MHz, and you wish to find the difference between them. Suppose further that you know each of the frequencies within an accuracy of 1%. For the sake of illustration, assume that the error of the higher frequency is on the high side and that of the lower frequency is on the low side. Here is what you would have:

$$204.5 + 1\% = 206.55$$
$$200.0 - 1\% = 198.00$$
$$\text{Difference} = \quad 8.55$$

This example shows that if there is a 1% error in each of the large numbers, there can be a 90% error in their difference.

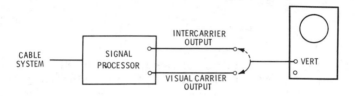

Fig. 15-6. Use of signal processor for frequency measurements.

It can be seen that the measurements must be as accurate as possible. The situation in practice is not likely to be as bad as the example for two reasons. First, the accuracy of a frequency counter is much greater than 1%, and second, the error will probably be in the same direction, either high or low, on both of the measurements. In our example, if both measurements were 1% high, we would have:

$$204.5 + 1\% = 206.55$$
$$200.0 + 1\% = 202.00$$
$$\text{Difference} = \quad 4.55$$

Now the error is only 1%, but it is still too high to meet FCC standards.

MEASURING SIGNAL LEVELS

The simplest way to measure signal levels is to use a signal-level meter. Inasmuch as most level measurements are made at subscribers' terminals, the fact that the signal-level meter is portable makes it the preferred instrument for these measurements. The technique is to connect the signal-level meter to the subscriber's terminal and measure the various visual and aural carriers of the signals carried by the system.

A quick way to measure all of the levels in a system is to use a spectrum analyzer as shown in Fig. 15-7. The measurement is made by noting both the height of the display on the screen and the setting of the input attenuator. Suppose, for example, that the height of a carrier on the screen indicates a level of −7 dBmV

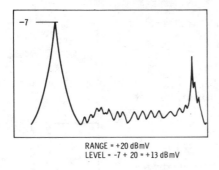

RANGE = +20 dBmV
LEVEL = -7 + 20 = +13 dBmV

Fig. 15-7. Measurement of signal level with spectrum analyzer.

and that the input attenuator of the analyzer introduces 20 dB of attenuation. The attenuation introduced by the attenuator must be added to the indication to get the true signal level. The correct signal level would be:

$$-7 \text{ dBmV} + 20 \text{ dB} = +13 \text{ dBmV}$$

Remember that if the spectrum analyzer is sweeping too fast, the indication will be in error. The way to avoid this is to reduce the sweep rate of the spectrum analyzer until the signals stop "growing taller." Further reduction of the sweep rate will contribute nothing to the accuracy of the measurement and may make the display harder to read.

When the sweep rate at which the spectrum analyzer is operating is harmonically related to one of the modulation frequencies in the signal, beats may appear on the screen. This usually appears in the form of a signal bobbing up and down on the screen. The sweep rate and dispersion must be adjusted so that the display holds still on the screen.

When a spectrum analyzer is used to make level measurements, the manufacturer's instructions should be consulted to determine how the measurement should be interpreted. In many spectrum analyzers, the accuracy reference is the top graticule of the screen rather than the bottom.

A part of the proof-of-performance measurement is a determination of the variation in signal levels over a 24-hour period. A convenient way of making this measurement is to connect a spectrum analyzer to the system and to photograph the display every hour or so for 24 hours.

FREQUENCY RESPONSE

For the proof of performance, the frequency response of each channel carried by the system must be measured. The response must be determined from the input at the antenna to the subscriber's terminal. In some systems, this is nearly impossible, so the practice in these cases is to find the frequency response of the headend and the remainder of the system separately. The two frequency-response curves for each channel are then combined algebraically to obtain the frequency response of the entire system.

One of the simplest methods of measuring frequency response is shown in Fig. 15-8. A sweeper is connected in place of the antenna at the headend, and a detector and oscilloscope are connected at the subscriber's terminal. The ramp waveform from the sweeper, which is normally connected to the horizontal input of the oscilloscope, cannot be used because of the long distance involved. Instead, a marker generator is set to place a marker on the signal at the low end of the range of frequencies being swept. This marker will provide a signal that can be used to trigger the oscilloscope. In this way, the low end of the frequency range being swept will be at the left-hand side of the oscilloscope display.

When the frequency-response measurement is made in this way, there will be no signal on the channel being tested, and the output of the sweeper will probably not be the same as the signal that is normally carried on the particular channel. This means that the agc in the signal processor at the headend will not be operating at

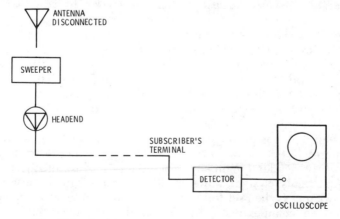

Fig. 15-8. Measurement of system frequency response.

the same level as it would be while handling a tv signal. This might very well render the frequency-response measurement meaningless. To avoid this problem either manual gain control should be used, or a dc signal should be fed to the agc of the signal processor so that it will operate at the same level of gain that it does while carrying the tv signal.

Although the method of making frequency-response measurements described here is fine for the regular proof-of-performance measurement, it has a limitation in that the regular signal must be removed from the channel while the measurement is being made. Thus, if this method were used as a regular maintenance check of the system, either it would have to be made at a time when there were no tv signals on the channel, or the subscribers would be subjected to a signal outage during the measurement. Other methods are available that will minimize subscriber inconvenience during the measurement. In general, these other methods should be used as a periodic check of frequency response, rather than as an absolute measurement for the proof of performance.

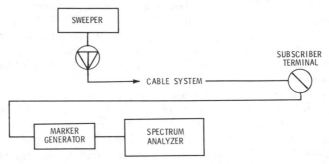

Fig. 15-9. Use of spectrum analyzer for quick check of frequency response.

Fig. 15-9 shows an arrangement that can be used to check frequency response with a minimum of outage time. The spectrum analyzer is connected to the subscriber's terminal in the same way that it would be connected for level measurements. The controls of the spectrum analyzer are set to display the channel on which the frequency-response measurement is to be made. The display should look something like that shown in Fig. 15-10.

The next step is to connect the sweeper into the headend. Initially, this step will take some time because it may be necessary to set the gain of the headend manually. Once the procedure has been refined for a particular system, however, it can be performed very quickly. The sweeper is then set for a very slow sweep rate, or even "rocked" across the channel manually. When this is done, the frequency response will be "painted" on the screen of the spectrum analyzer

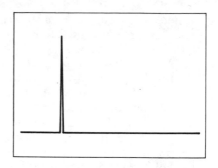

Fig. 15-10. Display during setup of
equipment for test in Fig. 15-9.

as shown in Fig. 15-11. Naturally, this method is much easier to use if the spectrum analyzer has a variable persistence or storage mode of operation.

With practice, this procedure can be performed by two technicians, one at the headend and one at the point where the measurement is to be made, in about ten seconds per channel. This outage period is so short that it will rarely result in subscriber complaints.

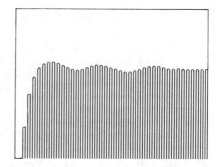

Fig. 15-11. Frequency-response
display obtained by method of
Fig. 15-9.

NOISE CALIBRATION OF A SIGNAL-LEVEL METER

Signal-to-noise measurements are made not only as a part of the proof of performance of a system, but also as a way to get the best possible performance from the system. Measurement of signal level has been discussed earlier. It is the measurement of noise level that is likely to be confusing. Many manufacturers of signal-level meters provide noise-calibration information that will permit accurate noise-level measurements to be made. Usually the FCC will accept measurements made with such calibration information. The information given here on how to calibrate a signal-level meter for noise measurements is presented because it shows how to make more accurate noise measurements, and it also gives some insight into the nature of noise and noise measurements.

There are three sources of error that are encountered when a signal-level meter is used to measure noise levels.

1. Noise power is proportional to bandwidth. The wider the bandwidth of any device, the more noise power will be present. The FCC rules specify noise level as the level of noise in a bandwidth of 4 MHz. Most signal-level meters have a bandwidth much less than this—on the order of 0.5 to 0.7 MHz. This will make the signal-level meter tend to read low when measuring noise levels. The amount of the correction factor (in decibels) that must be *added* to the indication to compensate for the bandwidth is:

$$\text{Correction factor} = 10 \log \frac{4}{\text{BW}}$$

where BW is the bandwidth of the signal-level meter in megahertz. For example, if a signal-level meter had a bandwidth of 600 kHz (0.6 MHz), the correction factor would be:

$$\text{Correction factor} = 10 \log \frac{4}{0.6} = 8.2 \text{ dB}$$

This means that 8.2 dB would have to be added to the indication of the signal-level meter to correct for the error due to the narrow bandwidth of the signal-level meter. Unfortunately, bandwidth is not the only source of error.

2. The FCC specifies noise in terms of its rms level. The detector of the signal-level meter, on the other hand, responds to the peak of the sync pulses of the video signal. Noise has a very high peak-to-rms ratio, so the characteristics of the detector of the signal-level meter will tend to make the indication high.

3. The detector of the signal-level meter tends to be more efficient at high levels, so it will read a higher value of noise at full scale than it would at, say, 10 dB down from full scale.

Of these three factors, the error due to bandwidth is usually greatest with the result that most signal-level meters indicate between 3 and 5 dB low when noise levels are measured.

In order to calibrate a signal-level meter to measure noise with some degree of accuracy, we need a noise source with a noise output that is at least 15 dB higher than the level of the noise generated inside the signal-level meter. A convenient noise source for calibrating a signal-level meter is a cable tv amplifier with a known noise factor and known gain. Unfortunately, manufacturers' published noise figures for anything except preamplifiers are likely to be worst-case figures, so it is necessary to measure the noise figure of the amplifier. A procedure for measuring amplifier noise figure is given in Chapter 17.

Fig. 15-12 shows a circuit for performing the noise calibration of a signal-level meter, using a standard amplifier as a noise source. The bandwidth of the amplifier must be much wider than 4 MHz, its noise figure must be between 6 and 10 dB and accurately known, and its gain must be about 40 to 50 dB. A standard amplifier of this type may be made of a distribution amplifier followed by a line amplifier, or two line amplifiers in tandem. The input resistor must be a 75-ohm coaxial terminating resistor.

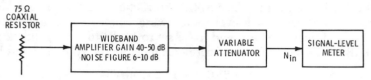

Fig. 15-12. Noise-calibration setup.

With the arrangement of Fig. 15-12, the noise level, N_{in}, in dBmV at the input of the signal-level meter will be given by:

$$N_{in} = N_o - A = F + G - 59.1 - A$$

where,

A is the attenuation of the variable attenuator in decibels,
F is the noise figure of the amplifier in decibels,
G is the gain of the amplifier in decibels,
N_o is the noise output power level from the amplifier in dBmV.

The value 59.1 is the thermal noise power in dBmV at room temperature, which was described in an earlier chapter.

Since the efficiency of the detector in the field signal-level meter is different at different parts of the scale, it is probably best to make noise measurements in such a way that the pointer of the signal-level meter will always be at the same point on the scale. This is accomplished by adjusting the variable attenuator for a given indication of the signal-level meter. The calibration procedure for the signal-level meter for this mode of operation is described below.

First, the gain and noise figure of the standard amplifier must be measured. The frequency at which the gain is measured is not critical as long as the amplifier gain is flat over at least 4 MHz. To calibrate the signal-level meter, tune it to the frequency at which the gain of the amplifier was measured, and adjust the variable attenuator until the pointer indicates 0 dB on the scale. The correction factor in dBmV for the meter is then:

$$\text{Correction factor} = N_{in} = F + G - 59.1 - A$$

where the letters have the same meaning as in the preceding equation.

For example, suppose that the standard amplifier had a gain of 45 dB and a noise figure of 8 dB, and that the setting of the attenuator that caused the indication to be 0 dB was 22.5 dB. The correction factor would then be:

$$\text{Correction factor} = 8 + 45 - 59.1 - 22.5 = -28.6 \text{ dBmV}$$

To make a measurement of noise level, connect the signal-level meter to the noise source through the variable attenuator, and adjust the attenuator until the indication of the signal-level meter is 0 dB. Then add the -28.6-dBmV correction factor. For example, suppose that the signal-level meter were connected to a cable tv system and the attenuator had to be set to 11 dB to get an indication of 0 dB on the meter. The correct noise level on the system would be:

$$11 - 28.6 = -17.6 \text{ dBmV}$$

This method of making noise-level measurements with a signal-level meter and a variable attenuator has the advantage of requiring only one correction factor for all measurements. It is possible to eliminate the attenuator and use the signal-level meter alone for noise-level measurements, but when this is done, a separate correction factor must be found for each part of the scale. These correction factors can be determined by using the arrangement of Fig. 15-12. As in the above procedure, the amplifier must have a known gain and noise figure. The procedure is to adjust the attenuator for various indications on the scale and compute a correction factor that will apply to each deflection.

Table 15-2. Correction Factors for Signal-Level Meter

Pointer Position on Scale	Indicated Noise Level (dBmV)	Attenuator Setting (dB)	True Input Noise Level (dBmV)	Correction Factor (dBmV)
-10	-40	27.5	-33.6	6.4
-5	-35	23.5	-29.6	5.4
0	-30	19.6	-25.7	4.3
$+5$	-25	15.0	-21.1	3.9
$+10$	-20	10.3	-16.4	3.6

For example, suppose that the standard amplifier again has a gain of 45 dB and a noise figure of 8 dB. The signal-level meter is set to its lowest range, say -30 dBmV. The attenuator is then adjusted for various scale indications, and a table showing the true noise input is prepared. Table 15-2 is typical.

After the calibration chart has been made, meter indications can be corrected simply by adding the correction factor to the meter indication. The factor to use is the one that applies to the pointer indication closest to the indication obtained when making the measurement. Note that the correction factor applies to the pointer indication, not to the scale of the signal-level meter in use. Thus, it is not necessary to make a separate table for each range of the signal-level meter.

As an example, assume that the meter we used to prepare Table 15-2 was connected to a cable tv system while on the −30-dBmV range, and the indication was +2, indicating that the noise level is −28 dBmV. The correction factor is the one for the nearest scale indication where we calibrated the meter, in this case 0 dB. The true noise level is then:

$$-28 + 4.3 = -23.7 \text{ dBmV}$$

MEASURING SYSTEM NOISE LEVEL

In order to measure the noise on a channel, there must not be any signal on the channel. To measure the noise level of the entire system, the signal is removed by disconnecting the antenna from its feed cable and connecting a 75-ohm terminating resistor in its place, as show in Fig. 15-13. The signal-level meter may be connected to the system either directly or through an attenuator. Either of the methods described in the preceding paragraphs may be used to make the measurement. In either case, the noise level being measured

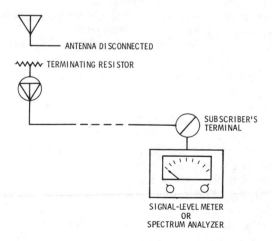

ANTENNA DISCONNECTED

TERMINATING RESISTOR

SUBSCRIBER'S TERMINAL

SIGNAL-LEVEL METER
OR
SPECTRUM ANALYZER

Fig. 15-13. Measurement of system noise level.

should be at least 15 dB greater than the noise generated by the signal-level meter. It is sometimes necessary to use a preamplifier ahead of the signal-level meter to get enough sensitivity to measure noise level in quiet parts of a system. When this is done, the preamplifier must be connected ahead of the signal-level meter while it is being calibrated so that the gain of the preamplifier will be taken into consideration in the correction factor.

If the manufacturer's correction data for noise measurements is available, this may be used for making the proof-of-performance measurements.

In any case, the gains of the amplifiers in the headend and along the system must be the same during the noise-level measurement as they are during normal system operation.

Some headend processors put a substitute carrier on a channel after the regular carrier has been off for more than some predetermined period of time, such as 20 seconds. Inasmuch as noise cannot be measured in the presence of a carrier, either the measurement must be made during the 20-second period when there is no carrier, or the substitute carrier must be disabled. This can often be accomplished by simply removing a crystal from the processor in the headend.

When noise is measured at a point in the system where the noise level is very low, say 55 dB below the adjacent-channel carriers, the signal-level meter may be overloaded by the adjacent-channel carriers. To avoid this, it may be necessary to use a bandpass channel filter ahead of the signal-level meter. If this is done, the same filter should be used ahead of the meter when carrier levels are measured for computing signal-to-noise ratio.

A spectrum analyzer can be used to get a good idea of the noise level in a system, and with proper correction of the indication, the spectrum analyzer can be used for actual noise measurements. The indication of the spectrum analyzer must be corrected for bandwidth errors and for the response of its detector, as was done with the signal-level meter. The procedure is the same.

The bandwidth of a spectrum analyzer can be confusing. If the instrument is set to display a wider range than a full tv channel, as in Fig. 15-14, there is a tendency to think that the bandwidth of the instrument is about 8 MHz. This is not true. The display is indeed 8 MHz wide, but it is made a little bit at a time as the spectrum analyzer sweeps from the low-frequency end to the high-frequency end of the display. The actual bandwidth of the analyzer is determined by the filter in its i-f stages. In most cases, the bandwidth is only a fraction of a megahertz. The correction that must be made for bandwidth is exactly the same as would be made for a signal-level meter; that is:

$$\text{Correction factor} = 10 \log \frac{4}{\text{BW}}$$

where BW is the bandwidth in megahertz of the filter in the spectrum analyzer. Suppose, for example, that a spectrum analyzer had a bandwidth of 0.2 MHz. The correction factor would be:

$$\text{Correction factor} = 10 \log \frac{4}{0.2} = 13 \text{ dB}$$

This means that the noise level displayed on the analyzer would be about 13 dB below the true noise level.

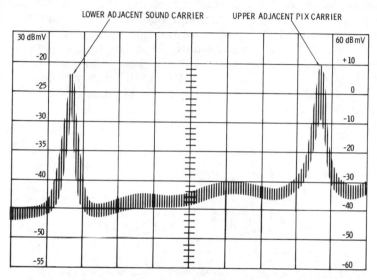

Fig. 15-14. Single-channel noise presentation on a spectrum analyzer.

The other correction that must be made is for the response of the detector in the spectrum analyzer. If the vertical deflection of the spectrum analyzer is in decibels, the detector is logarithmic and will tend to suppress the noise level. The amount of suppression is usually given in the manufacturer's literature and is typically 1.5 dB. Thus, the correction factor for our spectrum analyzer that happens to have a bandwidth of 0.2 MHz will be the sum of the two corrections, or about 14.5 dB.

All of the precautions that were listed above in connection with measuring system noise level with a signal-level meter also apply to use of the spectrum analyzer. If accurate measurements are required, the spectrum analyzer may be calibrated in the same way as a signal-level meter.

MEASUREMENT OF CO-CHANNEL INTERFERENCE

The same FCC specification that limits the signal-to-noise ratio on a cable tv system also specifies the minimum permissible ratio of signal to interfering co-channel signal. In a way, this is unfortunate because the measurement of co-channel interference is not the same as the measurement of system noise.

The best instrument for assessing the co-channel situation is the spectrum analyzer. Both the desired signal and any interfering signals can be seen on the analyzer display.

Before discussing what a co-channel signal looks like on a spectrum analyzer, let us look at a normal signal. Fig. 15-15 shows a visual signal as it appears on the spectrum analyzer with a resolution of 500 kHz or better. The display consists of the carrier together with a lot of little "carriers," or sidebands, on either side of the visual carrier. These sidebands are caused by the horizontal sync signal and are separated by about 15.75 kHz. These sidebands may be used as frequency markers to aid in identifying other signals that might be present in the channel.

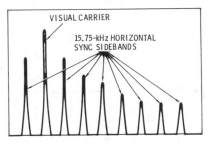

Fig. 15-15. Spectrum-analyzer display of normal visual signal.

The FCC regulations governing tv broadcast-station allocations are set up to minimize co-channel interference in home receivers. The basic protection against co-channel interference is geographical separation of co-channel stations. As an additional protection against co-channel signals that might possibly be picked up at the same location, the rules require that the carrier frequencies of some specified stations be offset 10 kHz from the nominal frequencies. For some of these stations, the offset is positive (higher frequencies); for others, it is negative (lower frequencies). Depending on which specific stations are involved, two co-channel signals could have carrier frequencies that are the same, 10 kHz apart, or 20 kHz apart.

When viewed on a spectrum analyzer, a co-channel carrier offset 10 kHz from the desired carrier will appear as an extra pip about two thirds of the way from the visual carrier to the first upper or lower sideband. This is shown in Fig. 15-16A. A carrier separated by 20 kHz will appear as a pip about 1⅓ sync intervals above or below the desired carrier. This is shown in Fig. 15-16B.

The displays shown in Fig. 15-16 are idealized somewhat in that there is no video in the display. If a small resolution and slow sweep

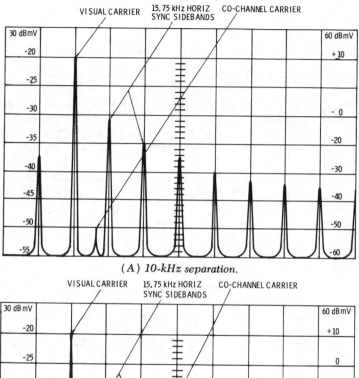

(A) 10-kHz separation.

(B) 20-kHz separation.

Fig. 15-16. Spectrum-analyzer displays for co-channel signals.

283

speed are used, the display will look enough like Fig. 15-16 to permit identifying co-channel carriers. Some spectrum analyzers have a video filter that makes the display much like that shown in Fig. 15-16.

The ratio, in decibels, of the visual carrier to the co-channel carrier can be found by counting the scale divisions between the peaks of the two carriers. In Figs. 15-16A and 15-16B, the co-channel carriers are about 30 dB below the regular visual carrier.

MEASUREMENT OF INTERMODULATION AND BEATS

Nonlinearity in a cable tv system will cause the regular signals carried by the system to heterodyne together and produce spurious beat frequencies. These spurious beats can be identified readily on a spectrum analyzer. Beats that are more than one megahertz from the visual carrier can be found by sweeping the spectrum analyzer across an entire channel; the resolution setting of the analyzer is not critical.

Fig. 15-17 shows a spectrum-analyzer display when there is a beat in the channel. Sometimes, beats that are actually below the noise level can be seen as a small "bump" in the noise. Any suspected beat should be watched to be sure that it is present for a period of at least several seconds. Often a video component of a picture will look like a beat but will disappear as soon as the program content changes.

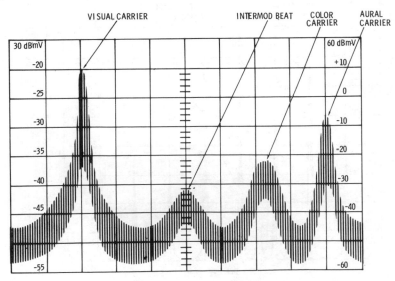

Fig. 15-17. In-channel beat.

When beats are close to the visual carrier, say within one megahertz, it is necessary to use a very low bandwidth and a slow sweep rate to be able to see them. The display under these conditions will show the horizontal-sync sidebands, and a beat will appear as an extra pip, as shown in Fig. 15-18. Any beat that is close to 10 or 20 kHz from the carrier should be viewed with suspicion, as it might well be a co-channel signal.

The ratio, in decibels, of visual carrier amplitude to beat amplitude is the difference between their heights on the spectrum-analyzer display.

Another technique that is often used to find spurious beats in the headend of a system involves only a signal-level meter. The signal-level meter is connected to the output of the combining network at the headend and tuned to the visual carrier frequency of the channel to be checked. The carrier is then removed from the system, and the signal-level meter is slowly tuned across the band to search for spurious signals. The sensitivity of the signal-level meter can be increased as required to find the spurious beats. This method suffers from a disadvantage in that it cannot detect beats that result from the visual carrier that is removed during the test.

When either a signal-level meter or a spectrum analyzer is used to measure beats, there is always the possibility that the spurious signal is actually being generated in the test instrument rather than in the system. This condition can be spotted by removing some of the attenuation in the attenuator of the instrument. The beat should

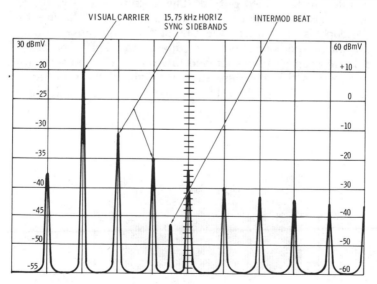

Fig. 15-18. Close-in beat.

change by the amount that the attenuator is changed. For example, if the attenuation is changed by 10 dB, the beat should change by 10 dB. If it changes more than this, it is probably being generated in the test equipment.

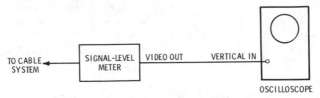

Fig. 15-19. Setup for measuring hum.

MEASUREMENT OF HUM AND LOW-FREQUENCY MODULATION

Hum and low-frequency modulation can be measured with a signal-level meter and an oscilloscope or with a spectrum analyzer. Some spectrum analyzers have a separate provision for measurement of hum. All that is required is to place a crystal-controlled modulated signal on the system and press a special button on the analyzer.

Fig. 15-19 shows a test setup for measuring hum and low-frequency modulation with a signal-level meter and an oscilloscope. To set up for the measurement, the signal-level meter is tuned to the carrier frequency of the channel to be tested, and the scope is dc-coupled to the video output. The attenuator of the signal-level meter is adjusted so that the level of the signal at the input of the oscilloscope is one volt to the tips of the sync pulses. The sweep rate of the oscilloscope should be adjusted to display one field of the signal. The scope is then switched to ac coupling, and the sensitivity is increased so that the difference in height of the sync pulses can be measured (Fig. 15-20). The voltage difference in the height of

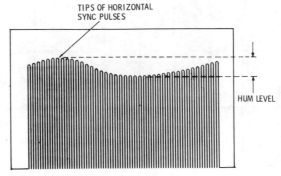

Fig. 15-20. Display of hum using arrangement of Fig. 15-19.

the sync pulses times 100 equals the percent of hum or low-frequency modulation on the signal. For example, if the difference in the height of the sync pulses on the oscilloscope display is 60 millivolts, this percentage of hum is $0.06 \times 100 = 6\%$.

TERMINAL ISOLATION MEASUREMENTS

The FCC rules require that the isolation between subscribers' terminals and the system be enough that open or shorted terminals at one subscriber's location will not interfere with picture quality at any other subscriber's terminal. The measurement of isolation is made with a signal source (which can be a sweeper that is capable of cw operation) and a signal-level meter. The procedure is shown in Fig. 15-21. First, the signal source is connected directly to the signal-level meter as shown in Fig. 15-21A. The level of the signal is adjusted to provide a convenient indication on the signal-level meter. The indication should be at the high end of the scale. Then the signal source is connected to one subscriber's terminal while the signal-level meter is connected to another subscriber's terminal without changing any of the controls on either the signal source or the signal-level meter (Fig. 15-21B). The amount of difference between the two indications is the amount of isolation between the two subscribers' terminals. Isolation measurements made in this way will cause interference unless they are made at frequencies where no signal is being carried.

An alternate technique that is sometimes used is shown in Fig. 15-22. Here the measurements are made at the ends of cables that

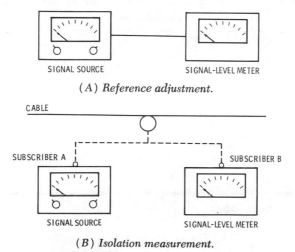

(A) Reference adjustment.

(B) Isolation measurement.

Fig. 15-21. Measurement of subscriber-terminal isolation.

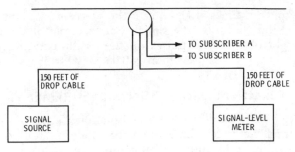

Fig. 15-22. Use of drop cable to measure tap isolation.

are typical of drop cables and that are connected directly to the splitter on the pole. This method gives about as good measurements as those made in the home, and it does not involve bothering the subscriber. Each cable can be terminated in a resistor and kept coiled up on the pole when measurements are not being made.

Superficially, it might appear that once terminal isolation has been established to be within limits, the isolation will not change unless the splitter is damaged. If the splitter were a resistive pad, this would indeed be true. Most subscribers' taps, however, obtain the required isolation because they work like directional couplers. As shown in Fig. 15-23A, any signal originating at a subscriber's terminal will be coupled to the line in the upstream direction and will not get into the next subscriber's terminal. Suppose, however, that there is a mismatch on the cable, as shown in Fig. 15-23B. Now any signal injected at subscriber A's terminal will travel upstream on the cable as shown by the solid line in the figure. At the point of

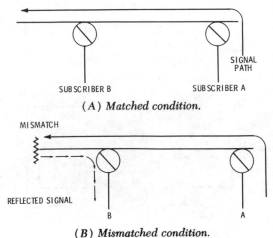

(A) Matched condition.

(B) Mismatched condition.

Fig. 15-23. Effect of mismatch on terminal isolation.

mismatch, a reflection will occur, and the reflected signal will be traveling in the proper direction to couple into subscriber B's terminal, as shown by the dashed line.

RADIATION MEASUREMENTS

Ideally, a cable tv system should not radiate any energy. Radiation is troublesome for many reasons. It may interfere with other radio services and lead to trouble with the FCC. Also, since radiated energy is energy that is being lost from the cable system, the signal-to-noise ratio is generally degraded. Finally, the ability of a system to radiate energy is closely related to its susceptibility to outside radiation; a system that radiates a great deal of energy will be susceptible to interference from outside radiation.

There are two ways in which a cable tv system can radiate energy. One possible source of radiation is the local oscillators in the headend. If there is not enough isolation between the local oscillator and the receiving antenna, some energy may be radiated out through the antenna. Another possible source of radiation, and this is the one that the rules seem to be aimed at, is the cable itself and its associated components such as connectors.

The primary purpose of radiation measurements should be to locate places where the system is radiating so that the radiation can be suppressed. For this purpose, the measurement setup need only be accurate enough to ensure that any residual radiation is well below the limits specified by the FCC.

Radiation is specified in terms of electromagnetic field intensity and is measured in units such as microvolts per meter. The first thing that is required to measure radiation is a device that will convert field intensity—which our regular instruments will not measure—to a voltage—which our regular instruments will measure. The device that does this is, of course, an antenna. The FCC rules suggest that a half-wave dipole antenna be used, and many manufacturers of signal-level meters and spectrum analyzers furnish antennas that can be used with their instruments to measure field intensity. In many cases, calibration curves are supplied so that the indication of a signal-level meter or a spectrum analyzer may be converted to field intensity in microvolts per meter.

We will first consider using a dipole antenna for measuring field intensities of signals above 25 MHz. Later, we will consider ways of measuring lower-frequency radiation. Fig. 15-24 shows a radiation test setup using a dipole antenna with a spectrum analyzer or signal-level meter. The arms of the dipole telescope so that the length of the dipole can be adjusted to resonance at the frequency at which a measurement is to be made.

289

The tuned dipole is a good antenna for radiation measurements because the relationship between the field intensity and the voltage at the terminals of the antenna can be computed without too much difficulty. The equation for the field intensity in terms of the voltage at the antenna terminals is:

$$E = 0.021fV$$

where,

E is the field intensity in microvolts per meter,
f is the frequency in megahertz,
V is the instrument indication in microvolts.

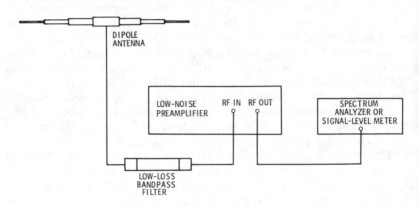

Fig. 15-24. Radiation test setup.

Since it is more convenient to work with logarithmic units (decibels and dBmV), we will convert the equation to logarithmic form:

$$E_{dB\mu V} = 20 \log (0.021f) + V_{dBmV} + 60$$

where,

$E_{dB\mu V}$ is the field-intensity level in decibels above one microvolt per meter,
V_{dBmV} is the instrument indication in dBmV.

This equation can easily be solved for the field intensity in microvolts per meter by using a scientific pocket calculator. However, to make things easier, the values of the expression 20 log 0.021f are given in Table 15-3 for all of the visual carrier frequencies normally used in cable tv systems.

Before going further, we should take note of the fact that the test setup in Fig. 15-24 shows a preamplifier. In many cases, the situation is complicated further by the fact that some antenna other than a tuned dipole is used for the measurement. These two factors can be taken into consideration by changing our equation to:

290

$$E_{dB\mu V} = 20 \log (0.021f) - G_1 - G_2 + 60 + V_{dBmV}$$

where,

G_1 is the gain of the amplifier in decibels,

G_2 is the gain of the antenna with a dipole as a reference (if the antenna is a tuned dipole, this term is zero),

V_{dBmV} is the indication of the instrument in dBmV.

Table 15-3. Value of 20 log (0.021f)*

Channel	Value of 20 log 0.021f (dB)	Channel	Value of 20 log 0.021f (dB)
T-7	−14.72	A	8.24
T-8	−10.18	B	8.66
T-9	− 7.22	C	9.05
T-10	− 5.01	D	9.43
T-11	− 3.25	E	9.79
T-12	− 1.79	F	10.14
T-13	− 0.54	G	10.47
		H	10.79
2	1.56	I	11.10
3	2.43	J	13.25
4	3.22	K	13.49
5	4.40	L	13.72
6	5.03	M	13.94
7	11.40	N	14.16
8	11.69	O	14.37
9	11.97	P	14.58
10	12.25	Q	14.78
11	12.51	R	14.98
12	12.76	S	15.17
13	13.01	T	15.36
		U	15.54
		V	15.72
		W	15.90

*For practical use, the value in the table should be rounded off to the nearest integral number of decibels.

The conversion from decibels above one microvolt per meter to microvolts per meter may be performed on a pocket calculator by using the following equation:

$$E_{\mu V/m} = 10^{E_{dB\mu V}/20}$$

where,

$E_{\mu V/m}$ is the field intensity in microvolts per meter,

$E_{dB\mu V}$ is the value obtained in the previous equation.

The conversion is also given in Table 15-4.

Usually, the cable used to connect the antenna to the preamplifier is short enough that its loss can be ignored. If the cable is long

Table 15-4. Conversion From dBμV to μV/m

dB μV	μV/m
0	1
10	3
15	6
20	10
25	18
30	32
35	56
40	100
45	178
50	316

enough to have significant loss, the amplifier gain (G_1) should be reduced by the amount of the cable loss. For example, suppose that in the setup of Fig. 15-24 the amplifier had a gain of 24 dB and the cable loss happened to be 2 dB. The quantity G_1 in the equation would be $24 - 2 = 22$ dB. If, in this example, a measurement made at channel 4 produced an indication of -1.2 dBmV, we could find the field intensity in microvolts per meter as follows. First, from Table 15-3, we find that the value of 20 log (0.021f) is 3.22. Substituting this value and the indication into the equation gives the level of the field intensity as:

$$E_{dB\mu V} = 3.2 + 60 - 22 - 1.2 = 40 \text{ dB}\mu V$$

From Table 15-4, we see that this corresponds to a field intensity of 100 $\mu V/m$.

The FCC rules specify that the radiation measurement be made with the dipole in a horizontal position at least 10 feet above the ground. Preferably, the dipole should be directly under the part of the cable system being tested, but if this would bring the dipole closer than 10 feet to the cable, the dipole may be moved to one side. The dipole should be rotated in the horizontal plane and the greatest indication taken as a measure of the radiation.

Certain precautions should be observed when field-intensity measurements are made. First of all, the antenna should be separated from any conducting object such as a metal post or a power line by at least 10 feet. Conducting objects that are close to an antenna will change its pattern and render the measurement meaningless. The second precaution that should be observed is to identify the radiation and be sure that it is actually coming from the cable system. A portable television set that can be operated from a battery supply is very helpful in identifying signals.

One of the difficulties in radiation measurements is that measurements cannot be made on frequencies that are used by local tv stations while the stations are broadcasting. This is particularly annoying because this type of radiation causes interference and is the subject of many complaints. The first step in checking on a complaint of interference with a local tv station is to check the radiation of the cable at other frequencies. This will give some idea of how much the system is radiating, and the cause of radiation could be located and the situation corrected without actually making measurements on the offending channel. Another obvious approach is to make radiation measurements late at night when local stations are not broadcasting.

In spite of other possible approaches, there will be times when it is necessary to measure the radiation from a cable system when a local station is broadcasting on the same frequency. The best approach in this case is to use a spectrum analyzer as the measuring instrument. The signal on the offending channel is removed from the cable, and a cw signal is substituted for it. The spectrum analyzer is set to its narrowest resolution. The frequency of the substitute cw signal is then changed enough so that it can be seen on the spectrum analyzer beside the local broadcast carrier. If the radiation from the system is low enough, it may be necessary to open a tap temporarily to get enough signal to set up the signal source and the spectrum analyzer.

Some relief from local broadcast signals can be obtained by taking advantage of the directional characteristics of the dipole antenna. Although the lobe of a dipole is broad, there is a fairly sharp null at the ends. Thus, if one end of the dipole is pointed in the direction of the local station, pickup of its signal will be minimized.

In many systems, measurement of radiation at frequencies below 25 MHz is not a problem simply because the system does not carry any signals below this frequency. When a system does carry signals below 25 MHz, measurement of radiation can be a problem. At 10 MHz, a tuned dipole would be nearly 50 feet long. This antenna on top of a 10-foot pole would be unwieldy, to say the least.

The best way to make radiation measurements at these lower frequencies is to use a field-intensity meter with a loop antenna that is designed for the purpose. If the expense of such an instrument is not justified, the services of a consultant or organization that specializes in radio interference measurements can be obtained to make the measurement. A sensitive communications receiver with a probe antenna can be used to locate sources of radiation at frequencies below the tv channels, although quantitative measurements cannot be made by this means.

CHAPTER 16

Troubleshooting and Component Testing

No man-made system will operate indefinitely without failure, and the cable tv system is no exception. A good preventive maintenance program will undoubtedly minimize the number of failures that result in disruption of service, but it will not eliminate them. In this chapter, we will discuss both troubleshooting of a cable tv system and testing of individual components. The subjects are grouped together because almost any troubleshooting procedure involves the testing of components to isolate the faulty unit. Detailed tests and adjustments of amplifiers are covered in a separate chapter because these tests are more commonly performed in the shop with different test equipment.

TROUBLESHOOTING PROCEDURES

The troubleshooting procedure starts whenever the engineer or technician has reason to suspect that there is trouble in the system. This suspicion of trouble can come from any of many different sources. The most common, and at the same time least reliable, indication of system problems is the subscriber complaint. The subscriber, not having technical training, has no way of discriminating between faults in his receiver, faults in the cable system, transmission difficulties at the station, and interference from other services. He definitely wants to believe that the nature of the trouble is one that will not result in a repair charge, and he is therefore inclined to complain to the cable system operator before he calls a tv serviceman. The subscriber complaint gains credibility rapidly as other subscribers begin to complain. The action that should be taken in response to a complaint from a single subscriber is debatable, but in general there are enough possible cable-system faults that will affect a single subscriber that it is not safe to ignore complaints, except possibly from chronic complainers.

The second indication that trouble might exist or be imminent on a cable tv system is a routine system measurement that indicates one or more of the system parameters has changed since the previous measurement. This is an indication that cannot be safely ignored.

Finally, the most obvious indication of system problems is something that can be seen on the monitor or instrumentation at the headend. No sign of trouble is welcome, but in this instance the trouble is almost always in one of the components at the headend, so the process of searching for the faulty component is simplified considerably.

DEFINING THE PROBLEM

The first step that should be taken before starting any systematic troubleshooting is to define the problem. It is easier to locate all of the faulty components that are contributing to improper performance if the trouble is specified in as much detail as practical. In some cases, the definition of the trouble is very simple. It might be simply, "no signal on Elm Street." In other cases, the definition is more elusive. Regardless of the nature of the symptom, troubles can usually be grouped into one of three categories:

1. Complete breakdown of part of the system. When this occurs, there is no signal on any channel at one or more points on the system. Faults of this type are usually the easiest to isolate.
2. Intermittent failures at one or more points on the system. Intermittent faults seem to have all of the legendary perversity of inanimate objects. Frequently, the fault will not manifest itself in the presence of the technician. This type of trouble is apt to be particularly time-consuming. The cable tv technician is at a definite disadvantage compared with the tv serviceman. Whereas the tv serviceman can take an intermittent tv set to the shop and wait for it to fail, the cable tv system must be left in place. In locating intermittent faults, the object is to minimize the troubleshooting time.
3. Marginal performance of all or part of the system. This is an insidious type of trouble. Subscribers complain that the picture quality is not good, but they are often indefinite about the

type of degradation. Unless the technician has a good set of measurements of the parameters of the system when it is operating properly, it is very difficult to tell just which components are faulty. It is very important to correct any marginal performance as early as possible, because neglecting marginal components can lead to a situation in which the performance of many components has degraded to the point where only a wholesale overhaul of the entire system will bring performance back to acceptable standards.

In addition to the categories described above, the troubleshooter should have a good description of the nature of the trouble. At the very least, the symptom should be described well enough that it can be recognized when seen on a monitor connected to some part of the system. In addition, the symptom, when possible, should be described in terms of some measurement. This may be a simple signal-level measurement, or even better, a sketch or photo of how the trouble looks on a spectrum analyzer. Particularly marginal faults usually are harder to distinguish on parts of the system where there is more signal level. If quantitative measurements are available, these faults are much easier to locate.

SYMPTOM ANALYSIS

A phase of troubleshooting that can save a great deal of time and should not be neglected is an analysis of the symptom as it appears on the subscriber's set or a monitor connected to the subscriber's terminal. The tv receiver gives a graphic display of the signal that is applied to its antenna terminals. Often, this display will give nearly as much information about the performance of a system as will measurements made with specialized test equipment. In general, the ability to diagnose trouble by looking at the picture is a skill that comes from experience. There are, however, some guidelines.

More often than not, complaints that involve only one channel are headend troubles. A common situation of this type is that subscribers at certain locations complain of poor or no color on a particular channel. The real trouble is that the signal on this particular channel is weaker than it should be. At the headend the signal is strong enough to lock in the color. At other points, such as the far ends of extender cables, the signal is too weak, particularly for older tv sets. A similar complaint arises when the carrier frequency has drifted excessively. Where the signal is strong enough, sets with good afc will have no serious problem. Where the signal is weaker, and sets have a marginal afc circuit, complaints will arise. Sometimes this trouble can be isolated merely by asking the complaining subscriber if he can correct the situation with the fine tuning control on his tv set.

Almost all troubles with the cable itself or with the passive components such as splitters will result in weak signals or no signal at all on some parts of the system and reflections on other parts of the system. A detailed analysis of the reflections will often help to pinpoint the location of the fault. Later in this chapter, we will review the subject of reflections and how they are related to cable-system faults and symptoms.

SAVING TROUBLESHOOTING TIME

The objective of troubleshooting is not only to locate and correct faults in the system, but to accomplish this in a minimum amount of time. Complaints are a serious matter in any system, but they are less likely to result in dissatisfaction and loss of revenue if they are corrected quickly. There is no substitute for experience in minimizing troubleshooting time, but even the beginner can reduce troubleshooting time if he attacks the job in a systematic manner. There is usually pressure from both management and the subscribers when there is trouble in a system. This pressure can lead to confusion and a great deal of wasted effort unless the approach is systematic.

There are two types of complaints that indicate where the troubleshooting procedure should start. One is the case in which all of the subscribers on the system have the same complaint. Here the trouble is almost always at the headend, or between the headend and the first subscriber. Equally obvious is the case in which only one or two subscribers experience the trouble. Here, the place to start is at the subscriber's terminal.

In other cases, the best place in the system to start the troubleshooting procedure is not so obvious. In such cases, there is what can be called an "equal likelihood" that the trouble may be in any of several different parts of the system. In such cases, time can almost always be saved by dividing the portion of the system that contains the fault into two smaller portions in which the faulty component is equally likely to be located. The trouble can then be isolated to one of these two portions of the system. This portion can then be further subdivided into two smaller portions. After it has been established that one of these smaller portions contains the faulty component, the subdivision process can be repeated until the faulty component is located.

A cable tv system in which many components are connected in tandem lends itself well to this "divide and conquer" technique. The method is illustrated in Fig. 16-1. Here, a section of a cable tv system containing five amplifiers and the associated cables is known to be

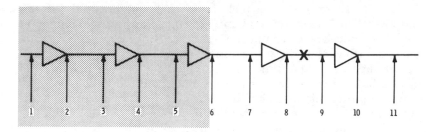

Fig. 16-1. Isolation of trouble by equal-likelihood division.

faulty. One way of looking for the trouble would be to start at the beginning of the section at point 1 in the figure and trace the signal through the system. A more satisfactory way would be to divide the section into two portions, each containing approximately the same number of components. For example, we could divide the system at point 6. The first step would be to check the signal at point 6. If the fault manifested itself at this point, we would know that the fault was in the shaded portion of the figure. Suppose that the signal were satisfactory at point 6. We then would know that all of the components in the shaded part of the figure were satisfactory, and we need waste no time checking them.

The next step is to divide the unshaded part of the figure into two smaller portions. For example, we could check the signal at point 9. Suppose that the signal is unsatisfactory at this point. The next step is to check the signal at point 8. Suppose we find it satisfactory there. We have isolated the fault to the section of cable between points 8 and 9 in the figure, and we had to make tests at only three points. If we had started at the beginning of the system, we would have made tests at nine places to isolate the trouble. Of course, this technique will not always save time. There is always a chance that a test made at a randomly chosen point will land right on the faulty component. This chance is slight, however, and in the long run the systematic approach will always save time.

In some cases, the preliminary isolation of the trouble may be done by telephone. If reliable viewers are available, they can often be contacted to verify whether or not the trouble is experienced at their location. This is particularly useful when the complaint is simply no signal. The installer will usually know which viewers can be counted on for a reliable signal report.

IDENTIFYING THE FAULTY COMPONENT

After the fault has been isolated to a particular part of the system, the next task is to identify the faulty component and repair or replace it. There are three techniques that can be used for this purpose. The first is a visual inspection. A surprising number of cable tv faults are mechanical in nature. Broken cables are common, as are damaged cases of amplifiers, passive components, and power supplies. Often, time can be saved by using a pair of field glasses to inspect the part of the system where the fault is located.

A second technique that will often save time is to correlate the symptom with some environmental phenomenon. If the trouble occurs during or after a heavy rain, a break in the integrity of the system (in either a cable or a component case) should be suspected immediately. If the fault manifests itself when the wind is blowing, connectors should be suspected.

Lastly, isolation of the faulty component can be accomplished either by measurements or by substitution of components that are known to be good. The advantage of the measurement approach is that the only instrument needed to locate many troubles is a signal-level meter or a spectrum analyzer. In smaller systems that use only a few different kinds of components, the trouble can often be located by simply substituting known good units such as amplifiers and passive components into the system. Of course, it usually is not practical to substitute coaxial cables; cable faults are usually best found by measurement.

The substitution approach to finding faulty components works very well when an amplifier is suspected. Amplifiers do not lend themselves to detailed testing in the field. When an amplifier is suspected of improper operation—particularly marginal performance—the best approach is to substitute a known good amplifier. If this clears up the trouble, the original amplifier is the cause of the trouble.

By substituting an amplifier that is known to be good into the various amplifier positions in the system, the amplifiers can be eliminated from the search. The trouble is then in either a power supply or the cable and its associated passive components.

Power-supply problems are likely to be subtle. They often occur when a surge causes a failure of service from the power line, but the complaints rarely come until after power has been restored to the neighborhood, because the subscriber cannot operate his tv set during a power outage. Sometimes, the utility people fail to restore power to a portion of the cable system after a power outage during which repairs were made to the power lines.

Once it has been established that the fault is either in the cable or in one or more of the passive components, it must be further isolated by testing. In the next few pages, we will review the subject of transmission and reflections in cables and passive components and how they affect the performance of the system.

ANOTHER LOOK AT TRANSMISSION
AND REFLECTION

The purpose of the coaxial cable is to transmit signals with a minimum amount of loss over the entire frequency band of interest. When there is a failure in a cable or a passive component, its transmission is reduced over all or part of the frequency range. In most cases, there are also reflections from the fault back along the system until they are blocked by an amplifier or a component with a great deal of loss.

Faults in cables and passive components are physical in nature. These faults are usually one of the following:

1. A conductor has broken.
2. A connection becomes loose or corroded.
3. Water has entered the cable or a component.
4. A conductor has moved in such a way as to cause a short circuit, open circuit, or impedance mismatch.

Each of these faults will manifest itself by changing the loss through the component or the reflection from the component, or both. This loss or reflection may occur at all frequencies or over only part of the frequency range.

Fig. 16-2 shows a section of coaxial cable. We discussed earlier the fact that if this cable were perfect, and if it were terminated in its characteristic impedance, there would be no reflection. Another way of saying the same thing is that the impedance seen looking into the cable would be a pure resistance, usually 75 ohms. Impedance-measuring equipment does not lend itself well to field use, so the fact that the impedance seen looking into a cable might disclose a fault is of little value in troubleshooting. We must specify faults in terms of quantities that we can conveniently measure in the field.

One quantity that we can conveniently measure in the field is signal level. For this reason, it is helpful to identify the ways in which various faults affect signal level or things that we can measure in terms of signal level. The first such quantity is the loss of a passive component or a section of cable. The loss of a section of cable is measured simply by measuring the signal level in dBmV at each end of the section and taking the difference in decibels. As was pointed out in an earlier chapter, this loss increases as the square

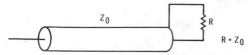

Fig. 16-2. Section of cable terminated for no reflection.

root of frequency, so it is necessary to measure the loss at more than one frequency.

Fig. 16-3 shows the method of measuring insertion loss. The signal level is first measured by disconnecting the input end of the cable and measuring the signal at the output of the amplifier. Measurements should be made at least at the visual carrier frequency of each channel carried by the system. The relationship among the strengths of the various signals at the output of the amplifier will depend on the type of tilt used in the system. The signal levels of the various channels are measured and recorded. Next the amplifier is reconnected to the input end of the cable, and the far end of the cable is disconnected from the system and the signal-level meter connected at this point. The loss in the cable at the various frequencies is then found by subtracting the indication found at the output of the cable from the signal that is applied to its input. This same general method of measuring the loss can be used with passive components.

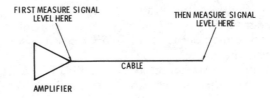

Fig. 16-3. Measurement of cable loss.

The second property of passive components that we can measure is the reflection. This measurement can be made in many ways, but the way that is best suited to field use is to define the reflection in terms of return loss. There are three reasons for this. First, manufacturers are increasingly specifying the reflection that can be expected from components in terms of return loss. This means that the measurement can be compared with the manufacturer's specification to verify the condition of the component. Second, the measurement can be made with a signal-level meter or a portable spectrum analyzer and a return-loss bridge. Finally, the return loss in decibels, when properly interpreted, gives a good idea of how much trouble the reflection will cause.

We stated in Chapter 5 that the return loss is given mathematically by the expression:

$$\text{Return loss} = -20 \log \rho$$

where ρ is the reflection coefficient, which is defined as the ratio of the voltage of the reflected wave to the voltage of the forward

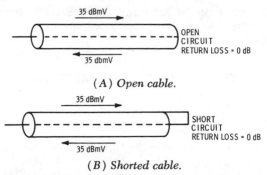

(A) Open cable.

(B) Shorted cable.

Fig. 16-4. Return loss of open and shorted cable.

wave. Looking at it another way, the return loss is the difference in decibels between the forward signal on the cable and the reflected signal caused by a mismatch.

Fig. 16-4 shows two coaxial cables, one terminated in an open circuit and the other terminated in a short circuit. Each cable is carrying a signal level of 35 dBmV. In each case, all of the forward signal is reflected from the end of the cable, so if we neglect the loss in the cable, the reflected signal will also have a level of 35 dBmV. The return loss from the termination is therefore 0 dB, meaning that the reflected signal has the same level as the forward signal.

In Fig. 16-5A, a cable is connected to a signal splitter that reflects very little of the signal. The forward signal has a level of 35 dBmV, and the reflected signal has a level of −2 dBmV. The return loss in this case is 37 dB, indicating that the reflected signal level is 37 dB below the forward signal. A reflected signal of this level is not likely to cause any noticeable picture degradation. In Fig. 16-5B, the same splitter has a fault so that it reflects about half of the incident voltage. The signal level is again 35 dBmV. The return loss in this case is 6 dB, indicating that the reflected signal will have a level 6 dB below that of the forward signal.

Expressing the level of the reflected signal in decibels puts the effect on picture quality in perspective. If we expressed the forward and reflected signals as voltages in the case of Fig. 16-5B, the for-

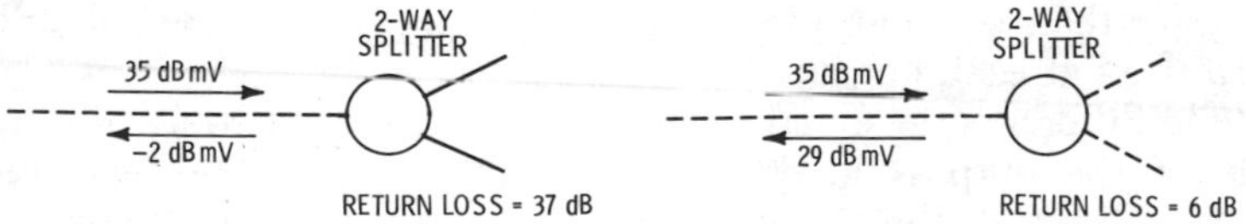

(A) Small reflection.   (B) Large reflection.

Fig. 16-5. Return loss from passive components.

ward signal would be 56 mV, and the reflected signal would be 28 mV. It must be remembered that a voltage ratio of 2 corresponds to a level change of 6 dB.

TESTING PASSIVE COMPONENTS

On the bench in the shop, testing of passive components is accomplished by using a return-loss bridge, a signal source, and sometimes a sweep generator. The tests are made to assure that the component meets the manufacturer's specifications. These tests should be made any time that a passive component is repaired. In the field, when a technician is trying to find out why his system does not operate properly, bench-type tests of components are much too time-consuming. The technician needs test procedures that can be accomplished quickly and will give him a general idea of whether or not the various components are operating properly.

The tests should be performed on at least the lowest and highest frequencies carried by the system. It is better to perform the tests on at least four channels.

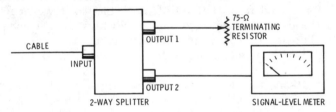

Fig. 16-6. Measurement of transmission loss through a signal splitter.

Fig. 16-6 shows a field test that can be performed to verify the operation of a splitter of the type that might be used on a drop cable to provide service to two subscribers. The first step is to disconnect the splitter completely and measure the signal level from the cable. The next step is to connect the splitter as shown in Fig. 16-6 and measure the transmission through the splitter to one of the outputs. The termination and the signal-level meter are then interchanged, and the signal level at the other output is measured. The levels at the two outputs of a good splitter should be the same. If more than two outputs are provided, the signal level should be measured at each output.

The difference in signal level between the input to a splitter and one of the outputs gives the loss that the signal experiences in passing through the splitter. This loss consists of two components: the loss that results from the division of the signal, and the internal losses in the splitter.

If a two-way splitter were ideal with no internal losses, the signal level at each of its two outputs would be 3 dB below the level of the signal at its input. This results from the fact that half of the power goes to one output and half of the other output. A power ratio of 2 represents a level change of 3 dB. (Do not confuse power ratios with voltage ratios when calculating level changes in decibels. A voltage ratio of 2 corresponds to a level change of 6 dB.)

In addition to the loss that results from splitting the signal, there will also be some internal loss in the splitter. In a good splitter, this is usually between 0.2 and 0.5 dB. Thus, a good splitter with two outputs would provide a signal level at each of its outputs that is about 3.2 to 3.5 dB below the signal level at its input.

In a four-way splitter, the loss resulting from splitting alone would be 6 dB; that is, the level at each of the outputs would be 6 dB below the level at the input.

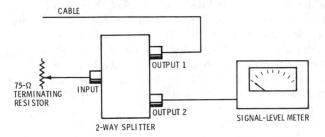

Fig. 16-7. Measurement of isolation between signal-splitter outputs.

The other test that should be performed on a splitter in the field is an isolation test. The setup for this test is shown in Fig. 16-7. Here the cable carrying the signal is connected to one of the outputs, the normal input is terminated in a resistance, and the signal-level meter is connected to the other output. The difference between the normal signal level from the cable without the splitter attached and the level measured with the arrangement of Fig. 16-7 is the isolation of the splitter. This value should be compared with the manufacturer's specification. A splitter that is water-soaked will rarely pass the isolation test.

Directional couplers and directional taps should be tested for insertion loss, isolation loss, and coupling loss. The insertion loss is measured by first measuring the signal level in the cable with the coupler disconnected and then making the measurement shown in Fig. 16-8. Here the output of the coupler is connected to the signal-level meter, and the tap is terminated in a resistance. The difference between the signal level in the cable and the signal level at the output of the coupler is the insertion loss.

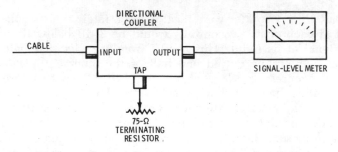

Fig. 16-8. Measurement of insertion loss of a directional coupler.

The isolation of the coupler is measured with the setup shown in Fig. 16-9. Here the cable is connected to the tap, the input is terminated in a resistance, and the signal-level meter measures the signal level at the output of the coupler. The difference between the signal level in the cable and the level measured at the output of the coupler is the isolation of the coupler. With good couplers, the signal level on the cable may not be high enough to result in a meaningful reading at the output of the coupler. In this case, a lab amplifier or an extender amplifier can be connected between the cable and the coupler. Of course, when this is done, the reference signal level is the signal level that is measured at the output of the amplifier rather than the signal level on the cable.

The last measurement to be made on a directional coupler is the tap loss. This is done by connecting the coupler as shown in Fig. 16-10. The difference between the signal level on the cable and the level measured at the tap is the tap loss. It should agree closely with the manufacturer's specification.

The tests described above can be made on just about any passive component in a system. They have two advantages: all of the tests can be made by using the signals on the cable as a signal source, and the only equipment required is a signal-level meter and one or more terminating resistors. If one of the components tests out marginally, it should be replaced and brought to the shop for a more detailed test.

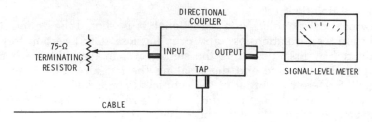

Fig. 16-9. Measurement of isolation of directional coupler.

The faults that are found in passive components usually fall into three categories. The most common fault is a defective connector. The connector is subject to stresses from thermal expansion and contraction, and from wind blowing on the cable. In addition, connectors are notoriously subject to corrosion. The next most common fault results from a break in the mechanical integrity of the unit. A small break in the case or around the connector will allow moisture to enter the unit and adversely affect its performance. The other common fault is an actual dent or mechanical bending of the component. Components that have been struck by tree branches in a storm or accidentally bumped by utility workmen often will not function properly. The mechanical shock displaces some of the internal components mechanically and thus disturbs their characteristic impedance at one or more frequencies.

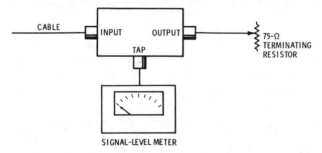

Fig. 16-10. Measurement of tap loss of directional coupler.

CABLE TESTING

Cable testing is complicated by the fact that it is necessary not only to determine whether or not a length of cable is faulty, but also to try to find the location of the fault. With a drop cable, the entire cable can and probably should be replaced whenever it has a fault. With trunk cables, particularly those with a large diameter, it is desirable to replace only the faulty section.

There are two steps that can save a lot of time in testing cables. The first is to check the connectors at the ends of the cable. Connectors are often faulty, and unless a bad connector has allowed moisture to enter the cable, replacing the connector may clear up the fault. The second time-saving step is to inspect the suspected section of cable visually with field glasses. Often the fault has an obvious cause—a dent or even a break in the cable.

Of course, there are cases in which the fault in a section of cable is not in the connector and is not obvious from a visual inspection. A signal-level meter can be used to verify that the cable is actually faulty. Comparing the loss in the cable with the manufacturer's specification will tell whether or not the cable is operating as it should. Measuring the dc resistance of the cable will give a rough idea of the distance to a short circuit, but this method will not work if there is capacitive isolation in the section.

If there is a short in the cable, resistance measurements made at each end will give a rough idea of where the short is located. In Fig. 16-11, R1 is higher than R2 because the short is nearer the end where R2 is measured. A careful visual inspection of the suspected region may show mechanical damage to the cable.

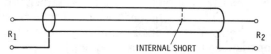

Fig. 16-11. Use of resistance measurements to locate short in cable.

Unfortunately, cable faults are not usually complete short or open circuits. More often, there is a partial short or a partial open somewhere in the cable. Uually, partial short circuits have more effect on the higher-frequency channels, and partial opens have a greater effect on the lower-frequency channels.

TIME-DOMAIN REFLECTOMETRY

The time-domain reflectometer is an excellent instrument for checking cables and passive components. Not only will it indicate that there is a fault, but it will also tell almost exactly where the fault is located.

The time-domain reflectometer (tdr), described in an earlier chapter, operates on the principle that an ideal cable has a constant resistive characteristic impedance throughout its length and will produce no reflections. If the cable has any anomalies along its length, these will cause reflections that will be displayed on the screen of the reflectometer.

If an ideal cable, terminated in a perfect resistance equal to its characteristic impedance, were connected to a tdr, the display would be as shown in Fig. 16-12A. An actual cable will provide a display more like that shown in Fig. 16-12B. The small fluctuations in the trace are caused by slight variations in the cable. These small reflections usually cause no problem because they are small and because they are distributed randomly along the cable so that the reflections will not add up at any particular channel frequencies.

A series loss in the cable caused by a high resistance in series with the cable will result in a slowly rising trace as shown in Fig. 16-12C. If the trace falls slowly, a shunt loss is indicated.

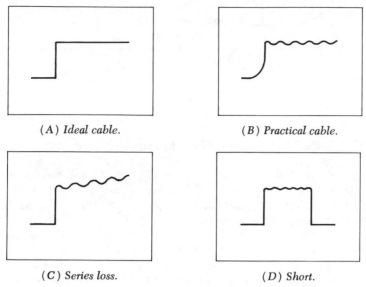

(A) Ideal cable.

(B) Practical cable.

(C) Series loss.

(D) Short.

Fig. 16-12. Displays on time-domain reflectometer.

The big advantage of the tdr is that it will show almost exactly where the fault is located. Fig. 16-12D shows the display that results from a short in a cable. The distance from the beginning of the cable is a direct function of the position of the discontinuity on the trace display. The tdr manufacturer's literature usually includes a conversion table that will enable the technician to tell within a foot or so where the fault is located. Another advantage of time-domain reflectometry is that it will show the condition of the entire cable. If the fault is such that moisture has entered the cable, the display will show how much of the cable has been affected.

FINAL WORDS ON TROUBLESHOOTING

The cable tv engineer or technician should not wait until trouble develops on a system before giving thought to troubleshooting. The test results obtained during routine proof-of-performance tests should be carefully recorded. When trouble develops, these test results can be used for comparison.

The more familiar a technician is with a system, the better he is able to interpret symptoms of trouble. Suppose, for example, that there is a complaint of no signal or very weak signal on the high channels to the left of subscriber A in Fig. 16-13. Suppose further that subscriber B complains of a severe ghost on the high-frequency channels, but plenty of signal strength. Immediately, the technician

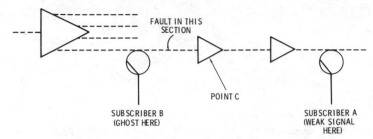

Fig. 16-13. Location of faults by analysis of reflections.

can forget all of the amplifiers between point C and subscriber A because the fault that causes the weak signal at point A is also causing the ghost at point B, and ghosts are not reflected back through the extender amplifiers.

Probably the most difficult aspect of all troubleshooting is breaking the news to a subscriber that the fault is in his tv set and not in the cable system. No one wants to hear that a fault of any kind is going to cost him money. The best way to break such news is to connect a picture monitor to the subscriber's terminal and show him what the picture would look like if his set were operating properly.

CHAPTER 17

Advanced Testing of Amplifiers and Passive Components

In the preceding chapter, we discussed methods of testing that could be accomplished in the field. These tests are usually made merely to decide whether or not it is advisable to replace a component in a system. When a component is removed from a system, the usual procedure is to take it to the shop for more advanced testing with more sophisticated test equipment. On the basis of these more advanced tests, a decision is made to discard the component, repair it, or return it to service. In this chapter, we will discuss some of the more advanced tests that are made on both amplifiers and passive components.

RETURN-LOSS TESTS

It cannot be overemphasized that the performance of a cable tv system is dependent on proper impedance matching of the components of the system. The better the impedance match between the various components, the better the performance of the system. One of the most convenient ways to express the impedance match in a cable tv component or amplifier is in terms of the return loss measured at its input port. Return loss should be measured at the video carrier frequency of each channel carried by the system, and preferably over the complete frequency range of interest. Sometimes

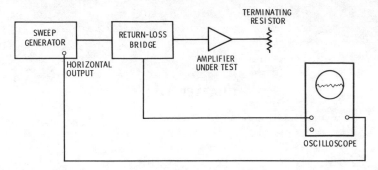

Fig. 17-1. Measurement of return loss.

an impending failure can be spotted from such measurements. A component may exhibit a satisfactory return-loss figure on all of the channels carried by the system, but may cause excessive reflections at other frequencies within its operating range. This is an indication that something is wrong, and although it might not disturb a particular system at present, the trouble will probably become more severe later.

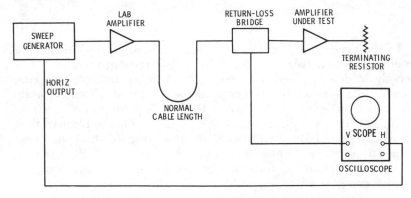

Fig. 17-2. Return-loss test with length of cable equal to system amplifier spacing.

A typical test setup for measuring return loss over a wide frequency range is shown in Fig. 17-1. Here the signal source is a sweep generator, and the measurement is made with a return-loss bridge. The return loss is displayed as a function of frequency on an oscilloscope. The component under test should be operated under the same conditions that it would experience when installed in a system. If a passive component is being tested, it should be operated with all of the ports except the input port terminated in 75-ohm terminating resistors. When an amplifier is being tested for return

312

loss at its input port, it should be operated as it would be when connected in a system. This means that the operating power should be furnished through one of the output connectors, and all of the outputs should be terminated with 75-ohm resistors. The gain controls should be set to their normal operating values, and if a pilot carrier is normally used, it should be used during the test. The signal level should be the same as would be encountered on the cable. This is sometimes accomplished by driving the amplifier through a length of cable equivalent to that used between amplifiers in the system, as shown in Fig. 17-2. Usually it is necessary to use a lab amplifier after the sweep generator to provide the proper signal level.

FREQUENCY-RESPONSE TESTS

A test that can be made conveniently in the shop is a test of the transmission loss or gain of a passive component or amplifier. A typical test setup is shown in Fig. 17-3. The signal source is a sweeper, and a detector is used to detect the signal, which is displayed on an oscilloscope. As in the return-loss tests, the component should be operated under the same conditions that it normally encounters when in service in a system.

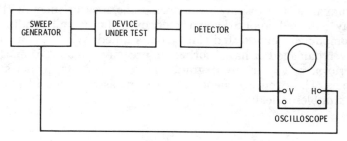

Fig. 17-3. Frequency-response test.

When an amplifier is being tested, it is sometimes desirable to use a cable equal in length to the amplifier spacing of the system, as shown in Fig. 17-4. When this arrangement is used, it is important that the output of the sweeper remain constant over the entire frequency range. If there is any doubt of this, the sweeper can be connected directly to the detector as shown by the dashed line in Fig. 17-4. After the connection is made, a grease pencil is used to mark over the trace on the screen of the oscilloscope (Fig. 17-5A). The connection is then made as shown by the solid lines in Fig. 17-4, and the grease-pencil line on the oscilloscope screen is used as a base line. The response of the unit under test is the difference be-

313

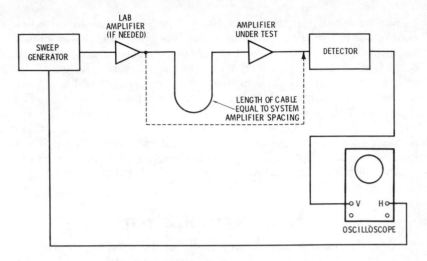

Fig. 17-4. Amplifier frequency-response test.

tween the actual trace on the oscilloscope and the grease-pencil base line, as shown in Fig. 17-5B.

Frequency-response measurements are useful because they provide a great deal of information about the component being tested. By comparing the display seen during the measurement with the display obtained from a known good unit, the overall condition of the component being tested can be determined. When a component has been damaged or has absorbed an excessive amount of moisture, its performance is usually degraded more at some frequencies than at others. The swept frequency-response measurement will show up these troubles quickly.

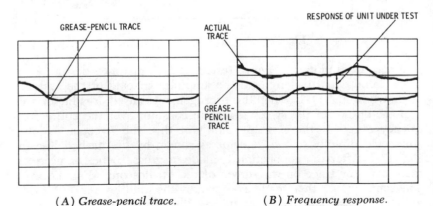

(A) Grease-pencil trace.   (B) Frequency response.

Fig. 17-5. Calibration of test setup in Fig. 17-4.

314

MISCELLANEOUS AMPLIFIER TESTS

The test setup shown in Fig. 17-4 for frequency-response measurements may also be used for many specific tests called for by a manufacturer of an amplifier. The sweep generator can be set for single-frequency operation and tuned, for example, to the frequency of a pilot carrier. The amplitude of the pilot carrier can be varied and the output signal monitored to see how the automatic gain control circuits of the amplifier are functioning.

The setup may also be used for adjusting traps and other amplifier circuits to meet the manufacturer's specifications. Inasmuch as special amplifier tests and adjustments vary considerably from one amplifier to another, the manufacturer's test instructions should be followed.

AMPLIFIER POWER-SUPPLY TESTS

The power supply in a cable tv amplifier performs three functions, each of which is essential to proper system performance. It rectifies the 30- or 60-volt ac signal that is usually transmitted along the cable and provides dc operating voltages for the amplifier circuits. It regulates the dc voltage so that the amplifier performance will not vary with variations in supply voltage. It filters the rectified voltage to reduce the ac ripple component to a level that will not interfere with amplifier performance. Each of these functions should be checked to assure that the power supply is operating properly.

Fig. 17-6 shows a test setup that will permit measuring the dc output voltage, voltage regulation, ripple, and noise from a power supply. The operating voltage for the test is derived from a regular cable tv power supply and is fed to the amplifier through the input signal connector. This voltage is varied by means of a variable autotransformer. The ac voltage at the input of the amplifier is measured with an iron-vane ac voltmeter. A regular ac voltmeter should not be used for this purpose because the waveform of the voltage from the cable tv power supply is not sinusoidal. The indication of the iron-vane voltmeter will be close to the true rms value of the voltage.

The dc operating voltage inside the amplifier is measured at a test point provided for the purpose by the manufacturer. This voltage measurement is made with a regular dc voltmeter, such as a multimeter.

The oscilloscope connected to the dc voltage test point in the amplifier is set to operate in the ac mode. This is so that it will not see the dc voltage and can be set to a sensitive enough range to permit measuring small levels of ripple and noise.

Voltage regulation is measured by varying the ac input voltage over the range prescribed by the manufacturer while noting how much the dc voltage changes. A typical manufacturer's specification might say that the dc voltage will not vary more than ±1 volt while the ac input voltage is varied between 18 and 30 volts as measured on the iron-vane ac voltmeter. While the voltage is being varied, the

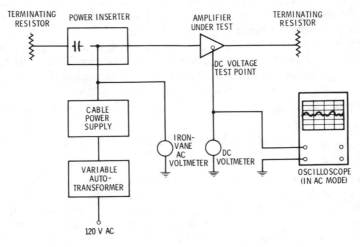

Fig. 17-6. Setup for power-supply measurements.

amplitude of the ripple and noise seen on the oscilloscope should be monitored to be sure that it does not exceed the manufacturer's specification.

MEASURING AMPLIFIER NOISE

The measurement of amplifier noise is becoming more important as amplifier design improves. In the early days of cable tv, all amplifiers were noisy, and there was little point in using highly refined methods of measurement. As better amplifiers become available, it is important to be able to determine that their noise performance has not deteriorated.

The noise performance of an amplifier should be checked whenever there is reason to suspect that the noise figure may have increased. This may be when the amplifier is removed from the system because it is noisy, or when the amplifier has been repaired in the shop.

Many noise-figure measuring sets are available that are priced within the reach of the cable tv system operator. Although all of these instruments are accompanied by complete instructions that are easy to follow, it is important for the engineer to have an appreciation of how the measurement is made.

Most noise-figure equipment makes the measurement through the use of a source of noise. It is not immediately obvious just how we can determine the quietness of an amplifier by putting noise into it. The principle can be explained with the use of a few equations, and in the process we will gain an insight into the subject of noise and noise measurement.

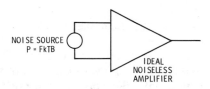

Fig. 17-7. Equivalent circuit of a practical amplifier.

In describing the noise performance of an amplifier, it is convenient to think of the amplifier as being perfectly noiseless, with a source of noise connected to its input as shown in Fig. 17-7. For convenience in explanation, we will consider the gain of the amplifier to be a ratio, or pure number, rather than a decibel quantity. In the same way, we will deal with the noise *factor* F, of the amplifier, which is a number, rather than the noise *figure*, which is expressed in decibels. By doing this, we can avoid the use of logarithms in the explanation.

The noise power output of the amplifier shown in Fig. 17-7 would be given by:

$$P_{no} = G \times FkTB$$

where,

P_{no} is the noise power output,
B is the bandwidth of the amplifier in hertz,
F is the noise factor,
G is the amplifier gain,
k is Boltzmann's constant,
T is the temperature in kelvins.

This is exactly the same noise output power we could expect from a practical amplifier having a noise factor F, a gain G, and its input connected to a 75-ohm resistor with no signal input.

Fig. 17-8 shows the same amplifier connected to a noise source that will provide a broad-band noise power of P_n watts per hertz of bandwidth. The power that this source will contribute to the amplifier will be equal to P_n times B. Because the noise generated by the source will not correlate with the noise generated by the amplifier, we can add the noise powers. Thus, the noise input power of the amplifier of Fig. 17-8 will be:

317

$$P_{ni} = P_n \times B + FkTB$$

and the noise power output of the amplifier will be G times this value.

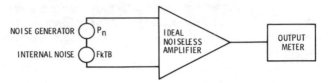

Fig. 17-8. Equivalent circuit of amplifier connected to noise generator.

Now suppose that we adjust the power from the noise source so that the noise power output from the amplifier is exactly twice what it was with the noise source not turned on. The output power will now be:

$$P_{no} = 2\ FkTB \times G$$

and the input power will be:

$$P_{ni} = 2\ FkTB$$

Inasmuch as the output power is simply the input power multiplied by the gain of the amplifier, we can write the following equation:

$$G[P_nB + FkTB] = G \times 2\ FkTB$$

It is immediately obvious that we can simplify this equation by cancelling both the amplifier gain, G, and the amplifier bandwidth, B, from both sides. This gives us the simpler equation:

$$P_n + FkT = 2\ FkT$$

We can rearrange this equation to read simply:

$$F = \frac{P_n}{kT}$$

What this equation tells us is that if we have a noise source and adjust it so the output power of the amplifier under test just doubles, we can calibrate an indicating instrument to show the noise factor of the amplifier. By simply marking the scale logarithmically, we can make the instrument indicate the noise figure in decibels. Most noise-figure measurement sets operate on this principle.

One of the advantages of this method is that it depends on neither the gain nor the bandwidth of the amplifier. The B that we used in the above equations is not the customary bandwidth of the amplifier, but the equivalent noise bandwidth, which will be somewhat different. This does not matter because the bandwidth cancelled out of the equations.

The accuracy of the method depends on how well we can know and control the output of the noise generator and how well we can detect the point at which the ouput of the amplifier under test has just doubled. The noise from the source should have the same characteristics as thermal noise. Some noise generators use special diodes for this purpose, and others use heated resistors.

MEASURING AMPLIFIER DISTORTION

In an earlier chapter, we discussed the fact that one of the factors that limits the number of amplifiers that can be cascaded in a cable tv system is the amount of distortion that each amplifier introduces. All amplifiers introduce some distortion. In general, the higher the output of a given amplifier and the more channels that it carries, the more serious the distortion.

One form of distortion called *second-order distortion* was taken into consideration when the vhf television channels were allocated. The channels were selected so that neither the second harmonic of one channel nor beats between two channels would fall within another channel. This means that a system carrying only the 12 vhf channels will not experience serious trouble from second-harmonic distortion. However, the midband and superband channels were not even dreamed of when the original tv channels were allocated, so systems carrying more than 12 channels must consider second-order distortion.

The other form of distortion that is significant is called *cross modulation*. This is the condition in which the picture information of one channel appears on the carrier of another channel.

There has been a great deal of confusion about just how distortion should be defined and measured. To resolve this difficulty, the National Cable Television Association (NCTA) has issued a standard covering distortion measurements. The NCTA standard recognizes two types of distortion—cross modulation and all other types of spurious signals. In the following paragraphs, we will first describe the principle of measurement; then we will consider ways of actually making the measurements.

Measurement of Cross Modulation

The principle of the measurement of cross modulation is illustrated in Fig. 17-9. The signal source provides a carrier at the visual carrier frequency of each of the channels to be carried by the amplifier. All carriers are adjusted to the same level. All carriers except the one on which cross modulation is to be measured are modulated

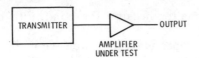

Fig. 17-9. Principle of cross-modulation measurement.

with a 15.75-kHz ±20% square wave. The resulting envelope for these channels is shown in Fig. 17-10.

If there were no cross modulation in the amplifier under test, the output of the amplifier *in the channel being tested* would appear as shown in Fig. 17-11A. If cross modulation were present, the waveform of the output of the amplifier being tested would be as shown in Fig. 17-11B. The cross modulation as a percentage is defined as:

$$\text{Cross modulation} = 100 \frac{a}{b}$$

where,

a is the variation in amplitude due to cross modulation by the 15.75-kHz modulation of the other carriers,

b is the amplitude of the unmodulated visual carrier with all other carriers turned off.

The objective of the measurement is to determine the value of parameters a and b in Fig. 17-11.

The measurement cannot be made directly with an oscilloscope because oscilloscopes that will respond to visual carrier frequencies are not available at most cable tv systems. Also, a method would be required to separate the other carriers from the carrier on the channel being tested.

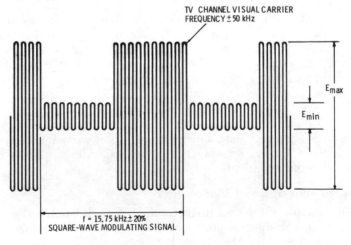

Fig. 17-10. Carrier modulation for cross-modulation measurement.

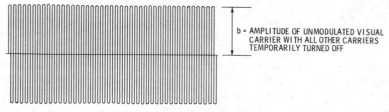

b = AMPLITUDE OF UNMODULATED VISUAL
CARRIER WITH ALL OTHER CARRIERS
TEMPORARILY TURNED OFF

(A) Carrier without cross modulation.

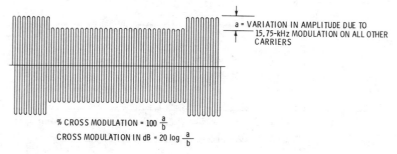

a = VARIATION IN AMPLITUDE DUE TO
15.75-kHz MODULATION ON ALL OTHER
CARRIERS

% CROSS MODULATION = 100 $\frac{a}{b}$

CROSS MODULATION IN dB = 20 log $\frac{a}{b}$

(B) Carrier with cross modulation.

Fig. 17-11. Output waveforms for cross-modulation measurement.

One method of making the measurement is shown in Fig. 17-12. Here the output of the amplifier is fed through a receiver having a bandwidth of about 400 kHz. An oscilloscope is connected to the detector in the receiver. First, the unmodulated carrier in the channel that is to be checked is fed to the amplifier with all other carriers switched off. The dc output of the detector displayed on the screen of the oscilloscope is a measure of parameter b (Fig. 17-13A). Then the other modulated carriers are switched on. The display will be as shown in Fig. 17-13B. The peak-to-peak value of the 15.75-kHz component is proportional to a. From these two parameters, the percentage of cross modulation can be calculated.

The object of the measurement is to measure the 15.75-kHz modulation that appears on what was originally the unmodulated carrier. A much more accurate and sensitive measurement of cross modulation can be made by using a receiver having a bandwidth of

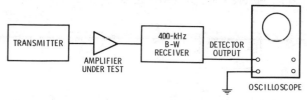

Fig. 17-12. Test setup for cross-modulation measurement.

321

31.5 kHz that will tune to the frequency of the unmodulated carrier. The receiver is connected as shown in Fig. 17-12, but the output of its detector is connected to an audio analyzer instead of an oscilloscope. An audio analyzer is similar to a spectrum analyzer that can be tuned to pass a single frequency (actually a very narrow band of frequencies). In this setup, the receiver will pass the carrier and the first sidebands. The audio analyzer will measure

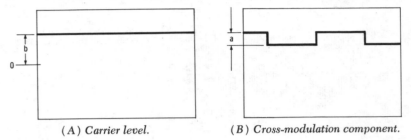

(A) Carrier level.   (B) Cross-modulation component.

Fig. 17-13. Waveforms for test of Fig. 17-12.

the amplitude of the 15.75-kHz component of the modulation. With this system, the unmodulated carrier is temporarily modulated 100% with the 15.75-kHz square wave, and the output of the audio analyzer is adjusted for a 100% indication. Then the modulation is removed, and the amplitude of the 15.75-kHz component that results from cross modulation is measured.

Measuring Spurious-Signal Ratio

The NCTA specification classifies all of the distortion-produced signals other than cross modulation as spurious signals. The measurement is called measurement of the spurious-signal ratio. The result can be expressed either as a ratio or in decibels.

The principle of the measurement is shown in Fig. 17-14. The object of the measurement is to supply to the amplifier being tested a series of unmodulated carriers at each of the visual carrier frequencies of the channels normally carried by the amplifier. The output measurement determines the ratio of the amplitude of the strongest spurious signal to the amplitude of the unmodulated carrier in the channel where the spurious signal appears.

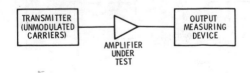

Fig. 17-14. Principle of measurement of spurious-signal ratio.

One way of making the measurement is with a spectrum analyzer. There are two factors that impose very stringent specifications on a spectrum analyzer used for this purpose. In the first place, the spurious signals may have frequencies very close to the carrier frequencies. This means that the spectrum analyzer must have a very high selectivity. Second, the spurious signals in a good amplifier may be as much as 90 dB below the amplitude of the unmodulated carriers. This means that the dynamic range required for the measurement is very large. The signal applied to the spectrum analyzer cannot be too large because the analyzer itself will then introduce spurious signals.

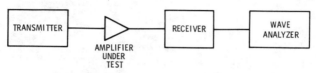

Fig. 17-15. Measurement of spurious signals.

Another method of measuring the spurious-signal ratio is shown in Fig. 17-15. Here the output of the amplifier is fed to a receiver having a bandwidth of from 1 MHz below the carrier to 4.2 MHz above the carrier. The output of the receiver is checked on a wave analyzer. The wave analyzer is similar to the audio analyzer mentioned earlier, but it has a frequency range greater than the audio range. The wave analyzer is tuned through the channel, and the amplitudes of the spurious components are measured. Care should be taken not to measure the harmonics of 60 Hz because these products usually result from insufficient filtering and not from non-linearity in the amplifier.

CHAPTER 18

System Integrity, Radiation, and Signal Ingress

A cable tv system can be thought of as a completely shielded enclosure with two sets of openings in the shielding. One set of openings consists of the headend antennas, where the desired signals enter the system, and the other set is the subscriber's terminals, where the signals leave the system. The performance of the system depends on handling the desired signals without intereference; therefore, it is important that any signals that belong outside the system stay outside. Likewise, other services use many of the frequencies carried by the cable system, and if interference to these services is to be avoided, signals that belong inside the cable system must stay inside the system (except where they are fed to the subscriber's receivers).

The ability of a system to confine signals and to avoid picking up extraneous signals is often called *system integrity*. Signals leaking out of a system are referred to as *radiation*, and signals leaking into the system are referred to as *signal ingress*. There are three places where signals can leak into or out of a cable tv system. The first is the headend. We want the desired signals to enter the system through the antennas at the headend, but there is always the possibility that an undesired signal will also enter the system here. Another obvious place where spurious signals may enter a system is at the subscriber's terminals. The purpose of the subscriber's terminal is to let signals leave the system, but it is possible that signals may also enter at this point. The third place where signals may unintentionally enter or leave the system is in all of the rest of the system. The system is intended to be a shielded enclosure, but frequently because of one fault or another the shielding is far from perfect. In this chapter, we will consider each of the places where signals may enter or leave the system with suggestions on how to pinpoint the problem and correct it.

For many reasons, the problem of interference to cable tv systems is becoming more serious as time passes. The tv channels for the most part carry only tv signals (except for a few areas where some of the channels may be shared with mobile radio or military services). The midband and super-band frequencies, however, are shared with many other services. Fig. 18-1 shows the services that use frequencies in this portion of the spectrum. Systems that use midband and super-band frequencies will generally experience more interference problems than will those systems that carry only the 12 vhf channels.

Another factor that aggravates the interference problem is the increasing use of the radio spectrum. As the number of transmitters increases, the number of possible interfering signals also increases. Particularly in the mobile service, an interfering source may be close to any part of the cable system. The number of transmitters in the Citizens Radio Service has increased drastically in recent years, and there are now several million transceivers in use. Most of these are in residential areas close to the cable system.

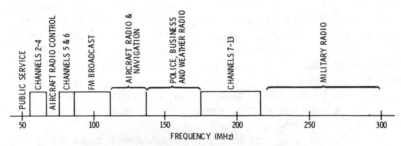

Fig. 18-1. Radio services in cable-tv frequency range.

Two-way cable systems that use frequencies below 30 MHz for upstream signals usually find that this portion of the system is much more susceptible to interference than the tv channels. The portion of the spectrum below 30 MHz is literally filled with signals, many of them from very-high-powered transmitters. When a system that was formerly free from interfering signals adds a two-way capability, interference is often one of the major problems.

THE EFFECT OF INTERFERENCE

One way that an interfering signal can affect picture quality is simply to come in "right on top of" the tv signal. This type of interference is serious, but usually it is also easy to identify. A more subtle type of interference is caused by intermodulation in the system. When a spurious signal enters a system, it can drive the amplifiers into the nonlinear region where heterodyning will take place. When two signals having frequencies f1 and f2 are heterodyned together, the resulting signals will have frequencies given by mf1 ± nf2, where m and n can be any integers. Thus, a single interfering signal beating with a single carrier in the system can cause spurious products in almost any part of the spectrum. This type of interference may affect any or all of the signals carried by the system and may be difficult to identify.

Interfering signals can be classified into three categories. The first is simply a signal from a transmitter on its assigned frequency. Such signals cause trouble by overloading components of the system and causing intermodulation products. The second class of interfering signal is a spurious signal from a transmitter. This is most commonly a harmonic of the carrier frequency, although parasitic signals are not uncommon. In either of these cases, the transmitter may or may not be operating within legal limits. Illegal transmitters are often difficult to locate. The operator usually is not particularly anxious to disclose his identity or location. Illegal linear amplifiers in the Citizens Radio Service are particularly likely to cause interference.

The final category of interference consists of what are usually called *incidental radiation devices*. These are devices that radiate rf energy incidental to their operation but are not intended primarily as radio transmitters. Examples range from induction heaters and diathermy devices to the ignition systems on snowmobiles that may frequent the vicinity of the headend in the winter.

THE RECIPROCITY PRINCIPLE

A theoretical principle that is sometimes useful in understanding interference problems is called the *reciprocity principle*. This principle states that any structure that will act as a receiving antenna will act equally well as a transmitting antenna. In practical terms, this means that if a cable system is susceptible to interfering signals, it is equally likely to radiate signals that may interfere with other services. Because of this principle, if a system is maintained so that it will not radiate signals, it will probably not be very susceptible to interference.

Another practical use of the reciprocity principle is that it is sometimes easier to find a leak in the system by radiating a small signal outside the cable and looking for the signal on the cable than to make the measurement outside the cable.

SHIELDING REVISITED

Most cases of radiation or signal ingress can be traced to faulty shielding. There are many misconceptions about shielding, and unless the basic principles are understood, shielding problems can be very perplexing. There is a tendency to think of a shield as protecting a circuit from interference in much the same way as a house protects its occupants from weather. A more correct explanation can be derived from basic field theory.

Whenever any electron in the universe moves at a given rate, it tries to make every other electron in the universe move at the same rate. This is why the electrons in a receiving antenna try to move at the same frequency as those in a transmitting antenna. The reason that a coaxial cable does not radiate is that the electrons in the outer conductor move in an equal and opposite manner to those in the inner conductor. Thus, the fields from the motion of the electrons cancel each other at the outer conductor. Only to the extent that the currents in the two conductors are equal and opposite will the fields cancel and the cable not radiate. This same principle applies to shielding. This is why a small break in a shielded enclosure will allow a system to both radiate and be susceptible to external signals.

Another principle that must be remembered when evaluating shielding is the skin effect. Because of the skin effect, all of the signals of interest in cable tv travel in an extremely thin layer on the surfaces of the conductors. Consider the situation shown in Fig. 18-2A. Here, both the cable feeding a piece of equipment and the equipment are grounded to a solid ground, as is the shielded enclosure. Superficially, it would appear that the arrangement would provide both good grounding and good shielding, but this does not consider the skin effect. Since the currents travel only on the surface of the conductors, the current path is as shown in Fig. 18-2B. The current flows from the inner conductor into the enclosure, through the equipment, along the inner surface of the enclosure, and back over the outer surface to the ground connection. The fact that the ground connection extends through the shield does not mean anything as far as radio frequencies are concerned. Far from providing shielding, the arrangement would act as an antenna.

The situation shown in Fig. 18-2 is an extreme case that would not be encountered in a properly installed cable tv system. It does, however, illustrate a principle that may be violated in many parts of a system. Calling to mind the skin effect and the way it affects a particular arrangement will often lead to the solution of a shielding problem.

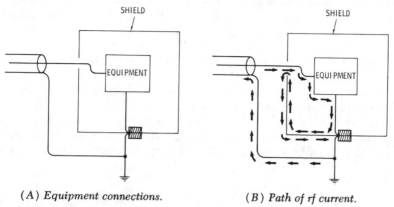

(A) Equipment connections.   (B) Path of rf current.

Fig. 18-2. Results of skin effect on shielding.

INTERFERENCE AT THE HEADEND

The antennas at the headend are designed to pick up the desired signals that are to be carried by the system. They may also pick up other signals that can cause interference. One type of interfering signal is a co-channel or adjacent-channel tv signal. This type of interference has been treated in an earlier chapter. Another type of interfering signal is one that has a frequency outside of the tv channels but is so strong that it gets into converters or preamplifiers and overloads them. Still another type of interference is a spurious signal from some transmitter. Educational fm stations radiating in channel 6 are examples of this type of problem. Harmonics of CB transmitters are also common.

If the signal that is causing interference is not in the tv bands, the problem can be solved with bandpass filters. The filters should be installed ahead of any active component such as a converter or preamplifier. Otherwise, overloading in the preamplifier or converter may introduce spurious products that cannot be removed from the system.

When the interfering signal is actually on the frequency of a tv signal, filters are to no avail. For example, if the second harmonic of a CB transmitter is within channel 2, no amount of filtering at the headend will remove it. The filtering must be done at the transmitter so that the harmonic will not be radiated. Before this can be done, the transmitter must be located. This is often the most difficult phase of the problem, particularly when the offending transmitter is in a mobile unit. Operators in the two-way mobile service are often negligent in identifying their stations. Illegal operators are careful not to identify their stations. The Field Operations Bureau of the FCC can sometimes help, but they tend to be overworked. Locating the offending transmitter often takes on cloak-and-dagger aspects with the cable tv engineer acting more like a detective than an engineer. Once the offending transmitter is identified and located, the operator will usually be cooperative. He is not anxious to pick up an FCC citation.

A very subtle type of interference that is becoming more common as the spectrum becomes more crowded is illustrated in Fig. 18-3. Here two transmitters are operating within legal limits with negligible spurious radiation. The signals from both transmitters combine in a metallic structure that has corroded to the extent that it

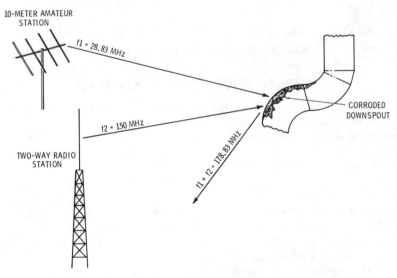

Fig. 18-3. Heterodyning of two signals in a corroded metal structure.

acts as a nonlinear device. Intermodulation occurs in the structure, and spurious signals are radiated. In some instances, the corroded structure may be a part of an antenna tower. In other cases it may be a downspout on a building that does not even have radio service of any type.

In the case shown in Fig. 18-3, a signal from a 10-meter amateur station beats with a signal from a 150-MHz two-way station to produce a beat that falls on the color subcarrier frequency of channel 7. This type of interference is troublesome for two reasons. In the first place, it is hard to locate, because both of the stations that are contributing to the interference are operating legally. Second, it is not easy to convince the owner of an item like a downspout that his downspout is causing television interference and must be repaired.

The last source of interference that must be considered is the incidental radiation device. If the device is in a fixed location, the FCC may be of help in getting the owner to clear up the interference. If the source is mobile, the problem is more difficult because the offending vehicle, such as a snowmobile, may well be long gone by the time the engineer gets to the headend. There is some hope, however, since the FCC is becoming concerned about the level of rf energy that is contributed by ignition systems. But until some other action is taken, this type of interference is best avoided by locating the headend where it will not be readily accessible to vehicles that may cause ignition interference.

INTERFERENCE FROM THE SUBSCRIBER'S TERMINAL

Another way that interfering signals can enter a cable system is through the subscriber's terminal. Although the balun transformer is usually connected directly to the antenna terminals of the subscriber's tv set, there is usually a short length of unshielded 300-ohm twin lead between the terminals and the tuner. This short length of twin lead is enough to pick up signals that can find their way back into the system. The subscriber is often unaware of this because the pickup will be effective only when his receiver is turned off.

A common source of interference at the subscriber's terminal is an illegal high-powered CB amplifier. A subscriber with a "linear" often finds that it interferes with his own tv reception, so he does not use his tv set when operating his transmitter. Although the offending signal is at 27 MHz, it may have harmonics in the tv channels, and it may find its way into an amplifier where it will beat with other signals and cause a number of spurious signals.

Yet another cause of signal ingress at the subscriber's terminal is a broken drop cable inside the home. Cables become broken with continual movement during housecleaning, and sometimes they are chewed by pets. Once the integrity of the cable is broken, it is vulnerable to interfering signals such as might be caused by vacuum cleaners or other appliances.

Interference entering through a subscriber's terminal is usually comparatively easy to locate if the engineer arrives in the vicinity while the interference is present. The interference is usually experienced only in the immediate vicinity of the subscriber's home. It cannot get back upstream through the downstream amplifiers, and often it is so weak that it is offensive only in the immediate vicinity.

SYSTEM INTEGRITY

By far the most difficult problem of signal ingress involves entry of a signal into the system through the cable itself or one of the shielded enclosures along the cable. A common place for interfering signals to enter the system is through the power supplies. The power lines carry both noise (from motors, neon signs, etc.) and rf signals that they pick up. Power supplies for cable tv systems contain filters that should reject both noise and spurious signals, but if the grounding and shielding are not correct, the filters will be ineffective.

Earlier, we stressed the principle that the effectiveness of shielding depends on the presence of equal and opposite currents in the shield and the conductors inside the shield. Even slight breaks in the shielding can destroy this action, and signals can enter and leave through the shielding itself. Most enclosures use mesh or conductive plastic gaskets to preserve the integrity of the shielding at places where doors or covers are located. If this material deteriorates, signals may enter or leave the system through a joint in the enclosure.

Corroded connectors are probably the most common offender as far as signal ingress is concerned. When interference is being traced, each connector as well as each component on the cable must be carefully checked.

INTERFERENCE TO UPSTREAM SIGNALS

The upstream channel in a two-way system is particularly susceptible to interference because of the lower frequencies used. This portion of the spectrum is used by just about every type of radio service, including broadcast stations. It is possible for the cable to pass through areas where the field intensity is greater than 50 volts per meter. In such extremely high field intensities, it is very difficult to obtain adequate shielding. If a 50-volt-per-meter field intensity is attenuated 100 dB by shielding, the field intensity inside the shield is still 500 microvolts per meter—a very high field intensity to have inside the shielded components of a cable system. Parts of a system within the extremely intense field of a broadcast station sometimes require tuned traps in the conductors themselves.

Fig. 18-4 shows a situation in which a section of cable and two ground leads actually form a loop which couples energy from a nearby broadcast station. It would appear that the current in this loop would be restricted to the outside of the cable. The current on the outside of the cable is indeed greater than that on the inside, but in a situation like this, enough current can flow inside the cable to cause serious problems.

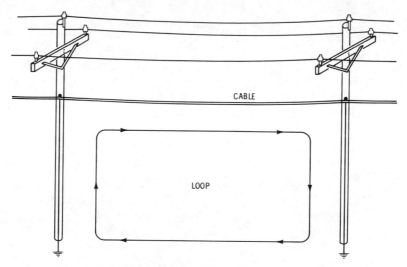

Fig. 18-4. Pickup loop formed by cable and two grounds.

One approach to a problem such as that illustrated in Fig. 18-4 is to install traps in the ground leads to reduce the current in the loop. Such an application of a trap is shown in Fig. 18-5.

Fig. 18-5. Detuning of ground lead to reduce rf currents.

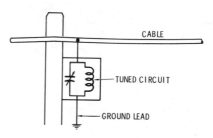

CHAPTER 19

Protection Against Lightning and Power-Line Surges

One of the most common causes of trouble in cable tv systems is failure of one or more components, or even the cable itself, due to excessive voltage or currents that enter the system from outside sources. There are three main causes of such damage:

1. Surges due to lightning. The lightning may or may not strike part of the cable system. Surges from nearby strokes can cause damage. These are high-voltage surges of very short duration.
2. Surges from power lines. These are due to line faults or load switching. They are not as strong as lightning surges, but they can cause a great deal of trouble.
3. Steady high currents in the outer conductor of the cable due to current on power lines. These currents are often not great enough to cause immediate damage, but they may cause components to break down from overheating.

PROTECTING THE ANTENNAS

The part of the system that is most susceptible to damage from lightning is the tall tower used to hold the antennas at the headend. There are three ways in which lightning can cause damage through the antennas. The most obvious is a direct hit; the lightning actually strikes one or more of the antennas. This can damage not only the antenna itself, but also the feed line and components in the headend. Solid-state devices such as preamplifiers and converters are particularly susceptible to lightning damage. The second way that lightning can cause trouble is for the stroke to hit the tower and not the antennas. Currents of thousands or sometimes millions of amperes can flow through the tower to ground. These enormous currents can induce very high voltages in the antennas and components mounted on the tower. A third and more subtle way in which lightning can cause trouble is for it to strike some nearby object; the electrical fields resulting from high current in nearby objects can induce voltages high enough to cause damage.

The situation that exists during a thunderstorm is roughly equivalent to that shown in Fig. 19-1. A thundercloud accumulates a large electrical charge. This charge concentrates on the bottom of the cloud and induces an equal and opposite charge on the surface of the earth beneath the cloud. When the difference of potential becomes great enough, the air ionizes and the result is a lightning stroke.

The conventional approach to lightning protection is the lightning rod. It has been known since the time of Benjamin Franklin that a pointed rod in a strong electrical field will start a discharge in the field. The antenna tower is protected by installing one or more pointed rods at the top of the tower, as shown in Fig. 19-2. These

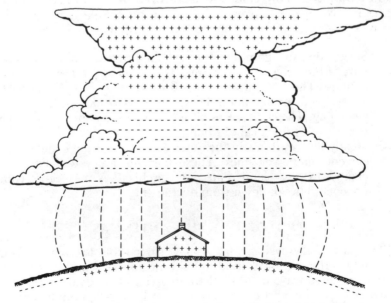

Fig. 19-1. Distribution of electrical charge during a thunderstorm.

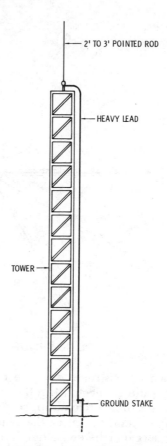

— 2' TO 3' POINTED ROD

—HEAVY LEAD

TOWER—

—GROUND STAKE

Fig. 19-2. Lightning-rod protection of
an antenna tower.

rods are bonded together, and a heavy conductor is connected from
them to a good ground. A good ground consists of several rods
driven in the earth to a depth of at least six feet. The concrete pier at
the base of a tower is not a good ground; lightning might find a
lower-resistance path to ground through the cable and one of the
headend components.

The purpose of the lightning rod is twofold. First, it drains off
static charges and thus reduces the likelihood of a lightning stroke.
(The capability of a lightning rod to eliminate strokes is a subject
that is hotly debated—as is almost everything about lightning.)
Second, when lightning does strike, it is guided safely to the earth,
with a reduced likelihood of damaging any components.

Recently a new method of lightning protection, or more correctly lightning elimination, has appeared. This scheme, illustrated
in Fig. 19-3, operates on the static drain principle. The system
consists of hundreds or even thousands of sharp points that are sup-

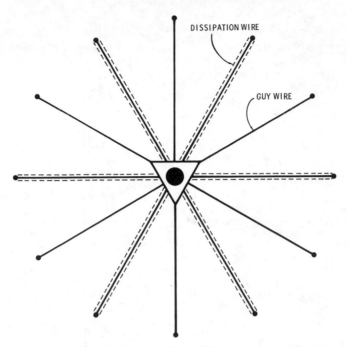

Fig. 19-3. Six dissipation wire radials strung from tower.

ported on a structure at the top of the tower. All of these sharp points are connected together to a wire leading to a good ground. When a thundercloud appears in the vicinity, the array acts like a high resistance between the cloud and ground, dissipating the charge on the cloud before a lightning stroke can occur. The equivalent circuit of the arrangement is shown in Fig. 19-4. The manufacturers of this system have reported that steady currents of over 100 amperes have been measured from such an array during a storm, indicating that it does drain the charge from clouds in the area.

In some instances, the damage caused by lightning is hard to detect. Connections in the coaxial feed cables may be damaged, or joints may become carbonized. Sometimes the semiconductor de-

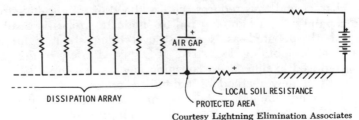

Fig. 19-4. Equivalent circuit of charge dissipation array.

vices in a preamplifier or converter become noisy without failing completely. Damage to the antennas and feed system can be detected by making vswr or impedance measurements. Whenever the vswr of an antenna increases for no apparent reason, the antenna and associated components should be checked thoroughly for possible lightning damage.

THE DISTRIBUTION SYSTEM

Although it is entirely possible for lightning to strike any of the components in the cable distribution system, this is not common where the cable system shares the same poles as power and telephone lines. The power lines are usually located much higher on the poles, and when lightning strikes, it usually strikes a component of the power system. There are three ways in which such a lightning stroke can affect the cable system. First, it can induce high voltages in the power system, and these can enter the cable tv system through the power supply. Second, very high currents are associated with a lightning stroke, and these can induce voltages in the cable system. Finally, ground currents can get into the cable system.

Fig. 19-5 shows a three-phase Y-connected power line that is on the same pole system as a cable tv system. Ideally, with all of the phases of the power system balanced, there is no current in the neutral line. This situation is rarely realized in practice. Steady neutral currents are experienced due to unbalance of the loads on the three phases. Very high neutral currents result from lightning

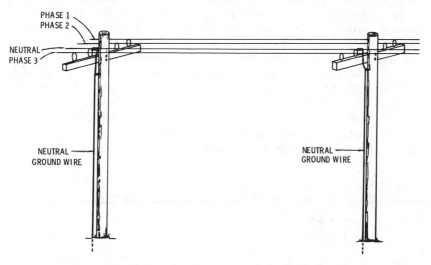

Fig. 19-5. Grounding of utility neutral wire.

surges and faults on the line. During a storm, wind may blow tree branches against one of the phases, or animals such as squirrels may short one of the phases to ground. These lightning and fault currents are of short duration, but they can cause a lot of trouble if they get into a cable tv system.

The neutral line of the power system is grounded at frequent intervals, and it would appear superficially that any current due to lightning would flow to ground through the nearest ground lead.

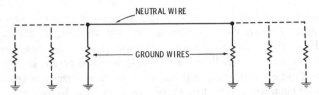

Fig. 19-6. Equivalent circuit of utility-system ground wires.

In fact, most of the ground current, but not all, will pass through the nearest ground leads. The entire system must be looked at as a group of resistances in parallel (Fig. 19-6). Those closest to the center of the diagram have the lowest resistance and will carry most of the current, but in a parallel resistance network, current passes through all of the branches. The currents in the higher-resistance branches will be smaller, but there will definitely be current in these branches. When the current has a magnitude of several hundred or even several thousand amperes, the small fraction that passes through the other branches will still be substantial.

It is not immediately obvious how these ground currents on a power system can get into a cable tv system. Fig. 19-7 shows a diagram of the equivalent circuit of a power-system neutral grounding system and a cable tv system running along the same pole line. A little study of this diagram will show that the outer conductor of the cable tv system is actually connected in parallel with the neutral line of the power system. In the figure, we have assumed that the resistance of the outer conductor of the cable is the same as the resistance of the neutral wire, which is often very close to the truth. If a large current flows through the neutral line of Fig. 19-7,

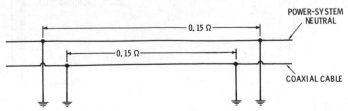

Fig. 19-7. Outer conductor of cable in parallel with power-system neutral wire.

340

a current of the same magnitude will flow in the outer conductor of the coaxial cable. The resistance of the outer conductor is small, so the voltage drop will be small, but nevertheless, there will be a voltage drop along the cable.

Remember that in addition to carrying the tv signals, the cable also carries 60-Hz power to the amplifiers on the system. If the voltage drop due to the ground current on the cable is in phase with the 60-Hz power being carried by the cable, this voltage will add to the supply voltage. High voltages that are of short duration will usually be taken care of by surge suppressors in the power supplies, and unless the current is great enough to damage the suppressor, they will not cause any trouble. Low voltages due to line unbalance, however, are not sufficient to trigger a surge suppressor. These voltages will simply add to the supply voltages and may cause excessive power dissipation in the amplifier power supplies. This type of high voltage is easily recognized because the supply voltage will be highest at the amplifier that is farthest from the power tap. This situation is a clue to the fact that voltages are being induced in the cable system by ground currents in the power system.

GROUNDS

All aerial cable tv systems must be grounded at regular intervals to minimize currents that may be induced by lightning and surges. What constitutes a good ground depends a great deal on the nature of the soil in the area. One common type of ground is an eight-foot rod driven into the earth. In other locations, it is necessary to use several rods connected together.

Usually, a power company has instruments for measuring the resistance of a ground, and if the power company has had any luck minimizing lightning problems, they will have a great deal of information on what constitutes a good ground in the particular area.

PROTECTIVE DEVICES

There are several types of surge-protection devices that are in common use in cable tv systems. About the only thing that can be connected directly across a cable is a spark gap. Almost anything else will have enough capacitance to disturb the impedance characteristics of the cable. At other points in the system, such as across the power supply in an amplifier, semiconductor devices may be used for surge protection. These devices behave like open circuits until the voltage exceeds a certain value, at which point they begin to conduct.

Surge protectors are almost always included by manufacturers of amplifiers and power supplies. These devices should be checked periodically because they may be damaged by one surge and thus be ineffective when the next surge occurs.

Graphic Symbols

SYMBOLS FOR USE IN CABLE TELEVISION*

POLES

Unless otherwise specified, the Standard considers wooden poles. Poles of other material, are designated by letters shown within or adjacent to pole symbols.

The Standard states that unless otherwise specified on drawings or referenced documents, pole usage and ownership are the same. For modification, the following letters are to be shown adjacent to the pole symbol.

Metallic	M
Concrete	C
Telephone	T
Joint	J
Power (Electric)	P

Pole ownership and conditions not covered above require appropriate designation within or adjacent to the circle.

The broken lines – ——— – indicate where line connection to a symbol is made and it is not part of the symbols.

Power Pole

Telephone Pole

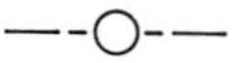

Joint Usage, Power and Telephone Pole

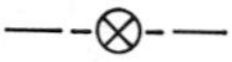

With Power Transformer

CATV Pole

Proposed Pole

or

Riser Pole
Use of a number indicates vertical distance from aerial connection to ground level.

Other Supporting Type Structures
Show approximate structure outline or a portion thereof in this manner.

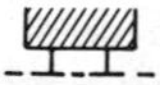

* *Courtesy National Cable Television Association*

Continued on Next Page

POLE LINE SUPPORT ELEMENTS In the following section, poles are shown as joint power and telephone poles because this occurs most frequently.	**Push Brace** To show a Push Brace, the support pole is a smaller symbol. It is proposed to be drawn in its actual supporting position relative to the pole it is bracing. **Extension Arm (Used by CATV)**	⊗⊗
CABLE SUPPORT ELEMENTS Stranded or Solid Messenger Items with poles shown are represented by:	**Tensioned Messenger Wire** **Slack Span, Messenger Wire** **Tensioned Messenger Wire Without Cable (Overhead Guy)**	
ANCHORING AND GUYING WITH POLES SHOWN	**Pole to Pole Guy** **Sidewalk Down Guy** **Sidewalk Down Guy with Anchor** **Down Guy** **Down Guy with Anchor** **Strut** **Tree Guy and Anchor** **Rock Guy and Anchor** **Building Guy and Anchor**	
UNDERGROUND DESIGNATIONS	**Underground Routing— Direct Buried** **Pedestal** Symbol on underground routing symbol, not cable symbol. **Manhole** See note under pedestal **Vault-Handhole** See note under pedestal **Underground Routing—Duct Line** The Standards for Underground Bore are only to be used if they are essential to show construction.	or (IEC) (IEC) or

344

	Direct Burial of Cable	
	Burial of Cable in Conduit	
	Underground Distribution System Marker	
HOUSE DROP DESIGNATIONS WITH POLES SHOWN	**Individual Dwelling** Numeral indicates the number of potential house drops at this location.	
	Multiple Dwelling Units Numeral indicates number of dwelling units, the arrow indicates the tap location.	15 UNITS
The asterisk is replaced by the proper designation.	**Non Residential Building (School, Church, Local Organization, etc.)** Designated appropriately as to the type of building or service.	*
MAKE READY OR POLE LINE PREPARATION REQUIREMENTS	Arrangement of existing equipment on the pole must be altered for CATV installation.	
POWER TRANSFORMER PLATFORM OR PAD		
AMPLIFIERS Again the broken line ___ _ ___ indicates line connection to a symbol and is not part of the symbol. Amplifiers on feeder lines (lines which may accept subscriber taps) should be drawn uniformly smaller than amplifiers on trunk lines. Asterisks are not part of the symbols. They are to be replaced with designations to denote the amplifier type. Designations will be explained on the drawing or referenced document.	**Amplifiers** (IEC)	
	Amplifier with subscriber distribution. (IEC) The dot indicates high output feeder line, if applicable.	
	Bridging amplifier with subscriber distribution (IEC)	
	Bridging amplifier with subscriber distribution and trunk termination.	
	Terminating (non-bridging) amplifier with subscriber distribution. (IEC)	
SPLITTING DEVICES	**2-way Splitter** (IEC)	
	3-way Splitter (IEC) Dot shows high output.	
	4-way Splitter (IEC)	

Continued on Next Page

	Directional Coupler (IEC) Model or value designations are to be shown adjacent to symbols. The high loss leg leaves from the angular half of the symbol.	
AC POWER BLOCKS	**AC Power Blocks** (IEC) A Two-way Splitter with AC Power Block shown as an example of application.	
EQUALIZERS The asterisk is to be replaced with a model number or performance characteristic (value) within or adjacent to the symbol and is not part of the symbol.	**Fixed Equalizer** (IEC) **Variable Equalizer** (IEC)	
SUBSCRIBER TAPS Note: The subscriber's port blocks AC power and the asterisk will be replaced with a model number or value.	**2 Output Directional Tap** **4 Output Directional Tap** **8 Output Directional Tap** **Other Taps** Taps with other numbers of drops and taps other than directional, e.g. pressure, should be designated by modification of these symbols or the listing of new symbols on the legend sheet for each design.	
FIXED FLAT ATTENUATORS The asterisk is replaced as on symbols for Equalizers and Taps.	(IEC)	
LINE TERMINATIONS	**(AC Power Blocking unless otherwise noted.)** or (IEC)	
POWERING DEVICES	**Cable AC Power Combiner** (IEC) **Power Supply** (IEC)	
CARRIER EQUIPMENT The Symbol points in the direction of its transmission. The asterisk is not part of the symbol and will be replaced with the appropriate descriptive designation.	**Carrier Control Generator** (IEC) Application: External	
	Application: Internal **Notch Filter (Narrow Band Stop Filter)** Application: External Application: Internal	

346

SIGNAL PROCESSING LOCATIONS (Antenna Site, Headend, Hub, etc.)	**Primary** Symbols are to be conspicu- ously large. *Headend* *Headend (Direct/* *(Aerial)* (IEC) *Origination)* (IEC) Local Distribution if Necessary Main Trunk **Secondary** (IEC) (IEC) Local Distribution if Necessary Main Trunk
COAXIAL CABLES NOTE: Specific coaxial cable function, other sizes, or type, should be designated by modification of these symbols or the listings of new symbols on the legend sheet for each design.	**.750 Inch (19.05 mm)** **Coaxial Cable** — — — — — **.500 Inch (12.70 mm)** **Coaxial Cable** — — — — — — **.412 Inch (10.46 mm)** **Coaxial Cable** ———————

Typical Coaxial
Cable Characteristics

Type	Dielectric	Attenuation (dB/100 ft, Ch 13)	Loop Resistance (Ohms/1000 ft)
RG-59U	Solid Polyethylene	5–6	57
0.412"	Expanded Polystyrene	1.4	1.6
0.412"	Expanded Polyethylene	1.6	2.3
0.500"	Expanded Polystyrene	1.1	1.2
0.500"	Expanded Polyethylene	1.3	1.4
0.750"	Expanded Polystyrene	0.71	0.5
0.750"	Expanded Polyethylene	0.95	0.67
1.000"	Air	0.55	0.27

Index

Aberration(s), 184
Accuracy, 266-267
Allocations, 119-121
Amplifier(s)
  bridging, 12, 102
  cable tv, practical, 102-105
  cascaded
    distortion in, 96-98
    noise in, 90-93
  mainline, 102
  specifications, 108-112
  tests, 315
  trunk, 12, 102
  two-way, 223
Angle, viewing, 185
Antenna(s), 123-136
  arrays, 131-135
  isotropic, 124
  log-periodic, 130-131
  pattern, 123, 125-129
    formation of, 132-135
  polarization, 128-129
  protecting, 335-339
  specifications, 135-136
  types, 129-131
  Yagi, 129
Astigmatism, 184
Attenuation, 77-79
  effect of, 200-202
  path, 234-235
Automatic operation, 112

Balun transformer, 13
Barrel distortion, 184
Baseband signal, 12
Beats, measurement of, 284-286
Boltzmann's constant, 22-23
Brightness, 31, 44
Broadside array, 131, 132

Cable(s)
  practical, 82-84, 86

Cable(s)—cont
  resistance, 170-172
  testing, 307-308
  trunk, 12
Camera, 178-186
  tubes, 178-183
Capture, 238
Carriers, pilot, generation of, 160
Channel(s)
  allocation, cable, 15-18
  classes of, 13-14
  twelve, more than, 212-215
Charge-coupled device, 182-183
Chrominance signals, 45, 46
Coaxial
  cable, 67-68
  devices, 84-86
Color(s)
  adaptation, 44-45
  perception of, 44-45
  primary, 44
  system, 45-48
  transmission of, 43-48
Compensation, temperature, 105-106
Components
  faulty, identifying, 299-300
  passive, testing, 304-307
Contrast, 32
Converter(s)
  down, 241
  set-top, 13
  signal, 148
  up, 239-240
Cross modulation, 60-61, 96, 319
  measurement of, 319-322
C-TAC, 120

dBmV, 21
Dc
  component, 42-43
  restorer, 43

349